Special Effect Glazes

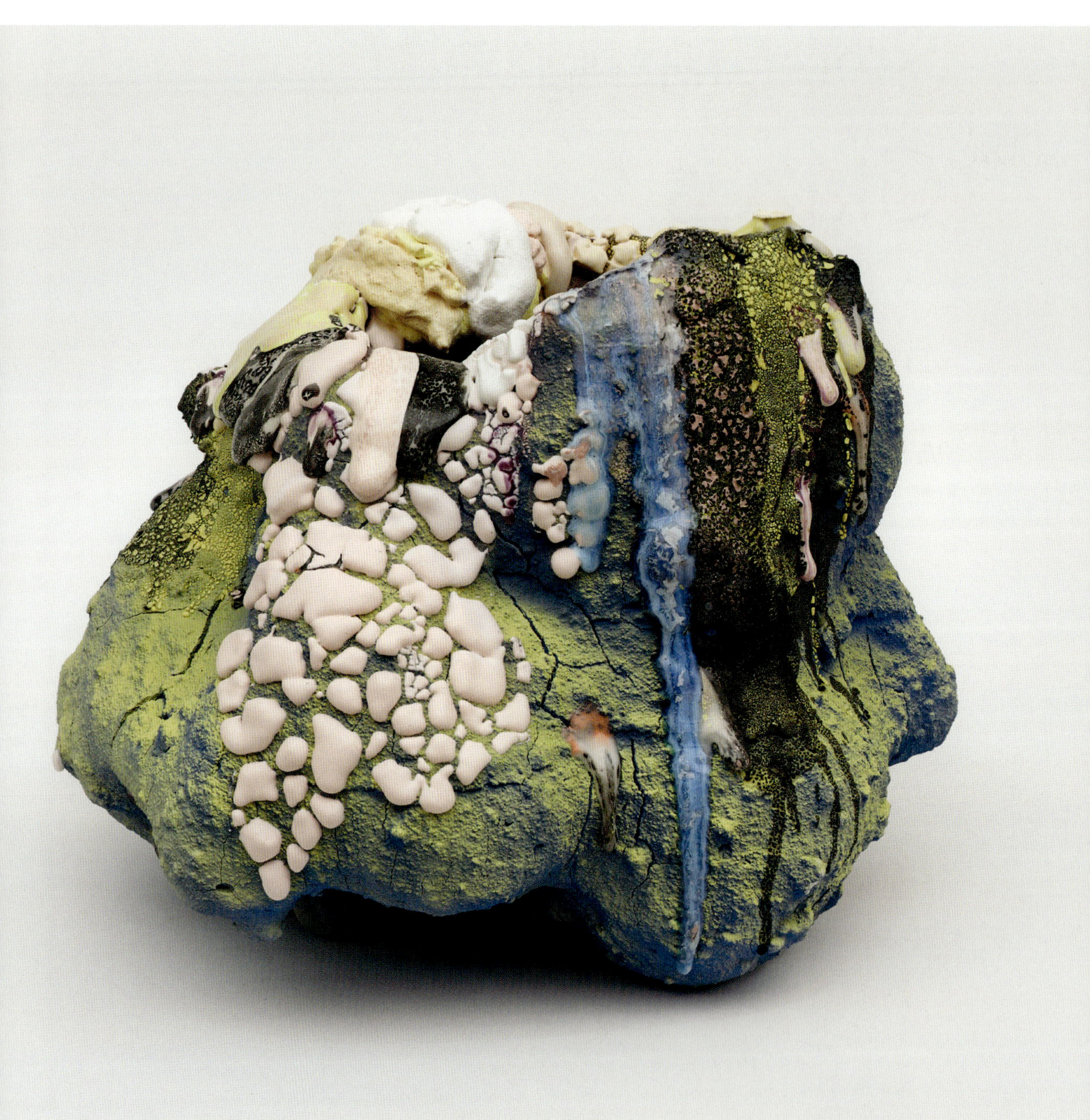

Special Effect Glazes

Linda Bloomfield

With photographs by Henry Bloomfield

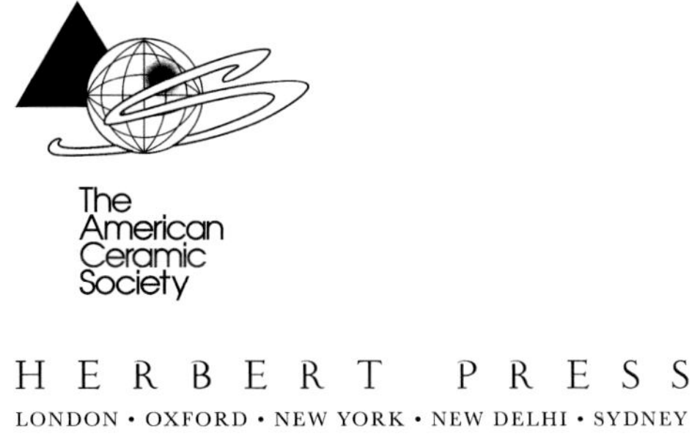

HERBERT PRESS
LONDON • OXFORD • NEW YORK • NEW DELHI • SYDNEY

HERBERT PRESS
Bloomsbury Publishing Plc
50 Bedford Square, London, WC1B 3DP, UK
29 Earlsfort Terrace, Dublin 2, Ireland

BLOOMSBURY, HERBERT PRESS and the Herbert Press logo are trademarks of Bloomsbury Publishing Plc

First published in Great Britain in 2020

Published simultaneously in the USA by
The American Ceramic Society
550 Polaris Parkway, Suite 510,
Westerville, Ohio 43082-7045

A catalogue record for this book is available from the British Library

UK ISBN: 978-1-912217-87-8
US ISBN: 978-1-574983-96-8

10 9

Designed and typeset in Rotis Semi Sans by Plum5 Limited
Edited by Alison Stace
Printed and bound in China by RR Donnelley Asia Printing Solutions Limited Company

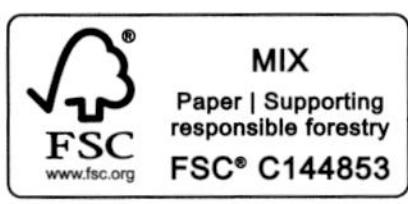

To find out more about our authors and books visit www.bloomsbury.com and sign up for our newsletters

FRONT AND BACK COVER: Brian Rochefort, *Cascado,* 2018. Stoneware, glaze, glass fragments, 43 x 40 x 48cm (17 x 16 x 19in.). Courtesy of the artist and Van Doren Waxter Gallery New York. *Photo: Marten Elder.* And Tessa Eastman, *Mint and Lollipop Baby Cloud Bundles,* 2017, height: 15cm (6in.). Private Collection. *Sylvain Deleu Photography.*

FRONTISPIECE: Brian Rochefort, *Cascado,* 2018. Stoneware, glaze, glass fragments, 43 x 40 x 48cm (17 x 16 x 19in.). Courtesy of the artist and Van Doren Waxter Gallery New York. *Photo: Marten Elder.*

Contents

Tessa Eastman, *Mint and Lollipop Baby Cloud Bundles*, 2017, height: 15cm (6in.). Private Collection. *Sylvain Deleu Photography*.

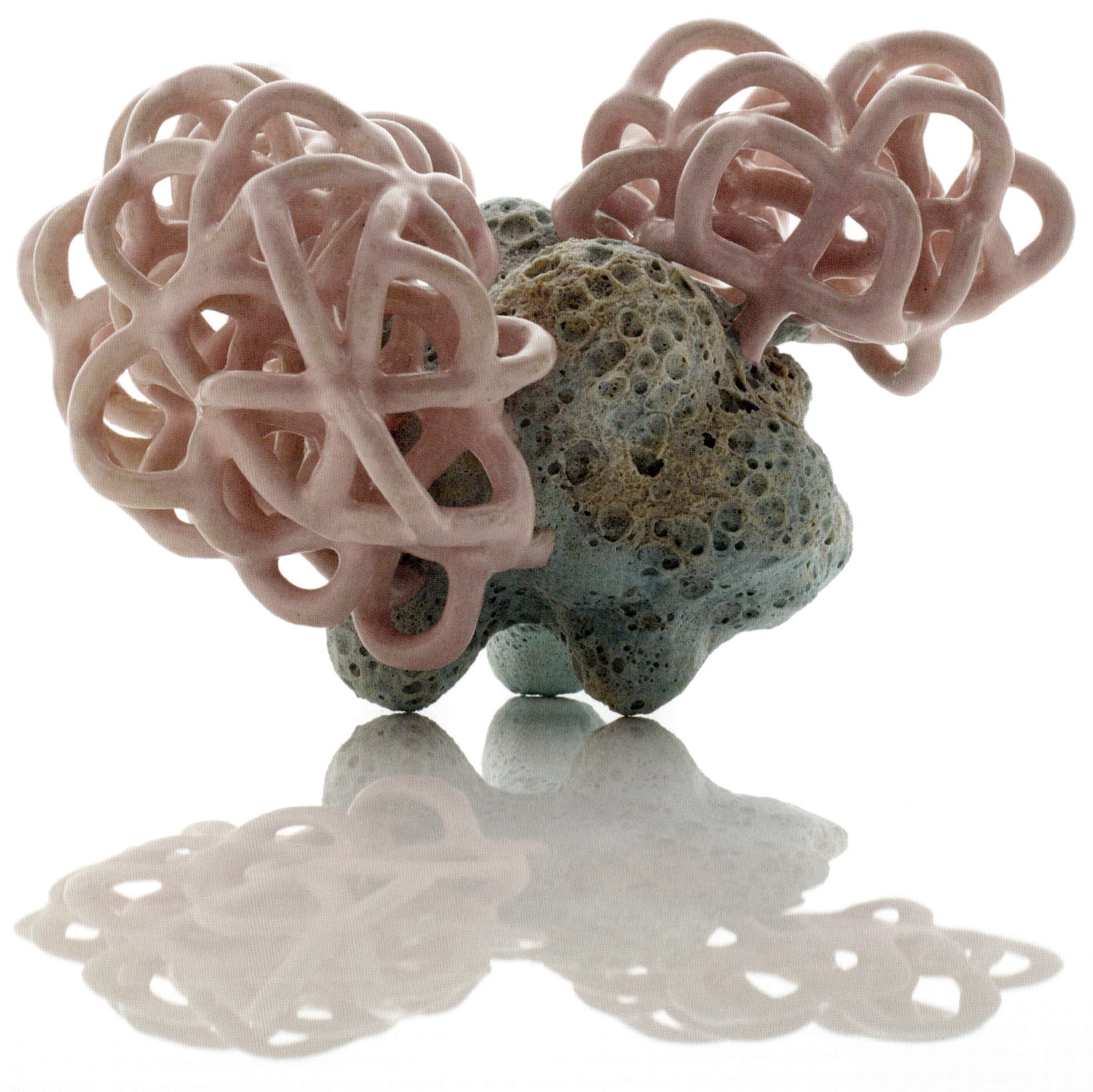

Acknowledgements

Thank you to Jayne Parsons at the Herbert Press and to Alison Stace.

Many thanks to Henry Bloomfield for photography, diagrams and proof-reading.

Thank you to Josefina Isaza for the volcanic glaze testing.

Many thanks to all the potters who contributed images and special thanks to Tessa Eastman for suggestions of potters using special-effect glazes.

Thanks to Matt Katz of Ceramic Materials Workshop for his research on zinc crawl glazes.

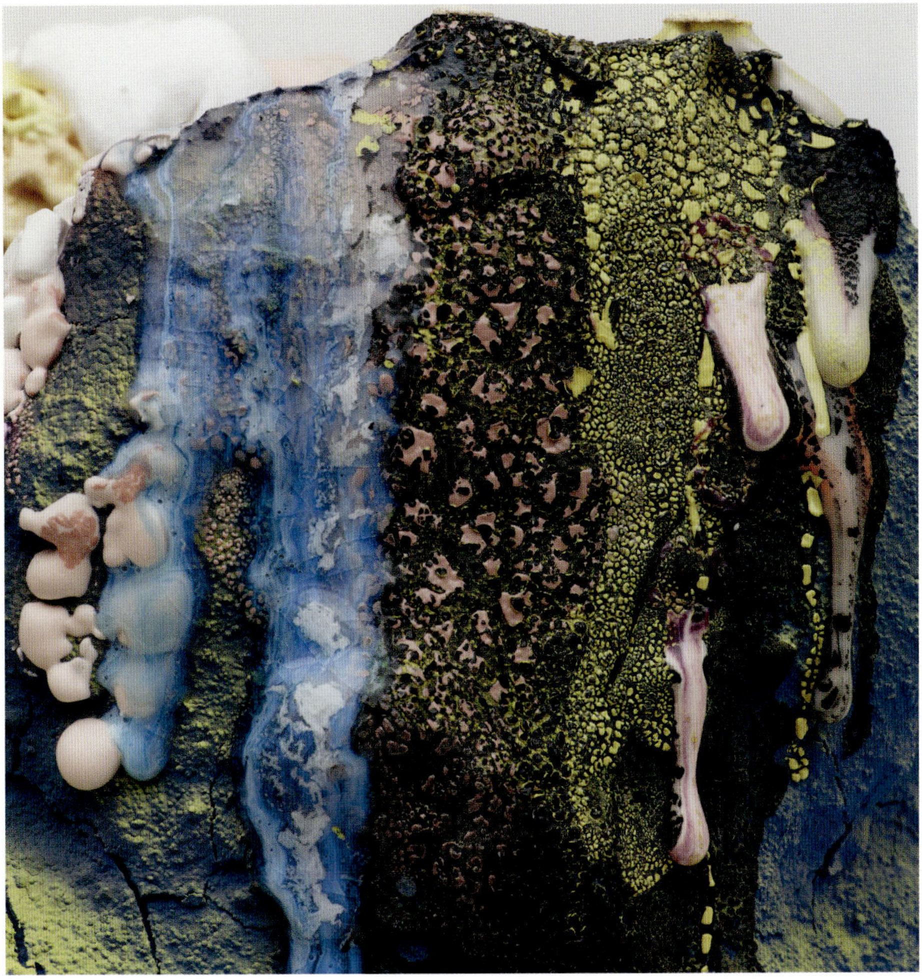

Brian Rochefort, *Cascado* (detail), 2018. Stoneware, glaze, glass fragments, 43 x 40 x 48cm (17 x 16 x 19in.). Courtesy of the artist and Van Doren Waxter Gallery New York. *Photo: Marten Elder.*

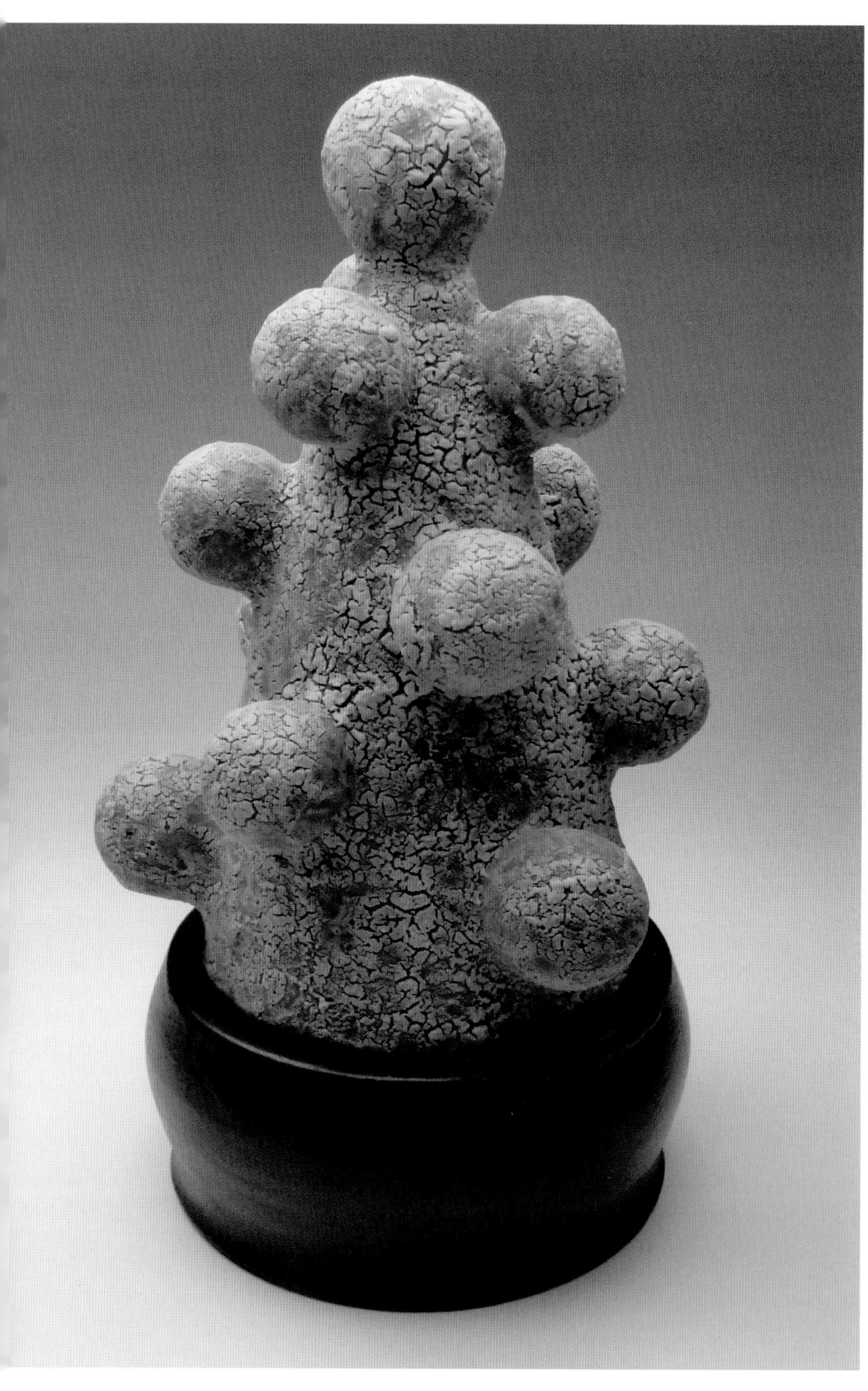

Virginia Scotchie, *Knob Cone Form,* 2018. Stoneware clay, wheel-thrown and handbuilt, textured glaze, mid-range-fired, 30 x 20 x 20cm (12 x 8 x 8in.).

Introduction

This book is not just a recipe book, but also sets out to explain the crucial principles of making special-effect glazes. These glazes are not wildly different from ordinary glazes, but are often outside the boundaries defining defect-free, glossy glazes. Many special effects are forms of glaze defect; crazing, crawling, pinholes and blisters can produce special-effect glazes known as crackle, lichen and lava glazes. Once you discover the techniques used to create these effects, you can experiment and make your own unique glazes.

Industrial tableware glazes are often transparent, glossy and well behaved, with no glaze defects such as crazing or crawling. However, many studio potters like to use glaze defects as interesting special effects.

Making glazes can seem daunting, but mixing a glaze is actually no more difficult than making a cake. Potters usually follow glaze recipes they find in books or online. The ingredients come as finely ground powders: quartz, whiting, feldspar and clay. The powdered materials are weighed out carefully and added to a container half full of water, then left to slake, until the water has completely saturated the powder. The glaze can then be sieved through an 80 mesh potters' sieve. Once the water content has been adjusted until the glaze consistency is somewhere between milk and single cream, either by adding water or leaving overnight to settle and removing water, then it is ready to use.

Like cooking, it is possible to make a basic glaze without much knowledge. It is important to measure the glaze ingredients very carefully with accurate scales and to test it first. Make a note of the recipe and number the corresponding test tile. However, if you can really gain an understanding of the science behind glazes, you can have much more control over what comes out of the kiln.

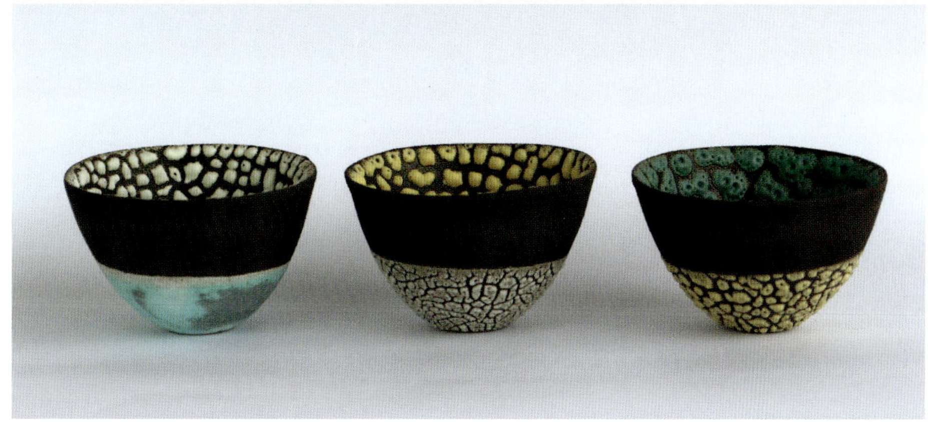

Emma Williams, press-moulded black earthenware clay, barium and crawl glazes, height: 8cm (3in.). Photo courtesy of the artist.

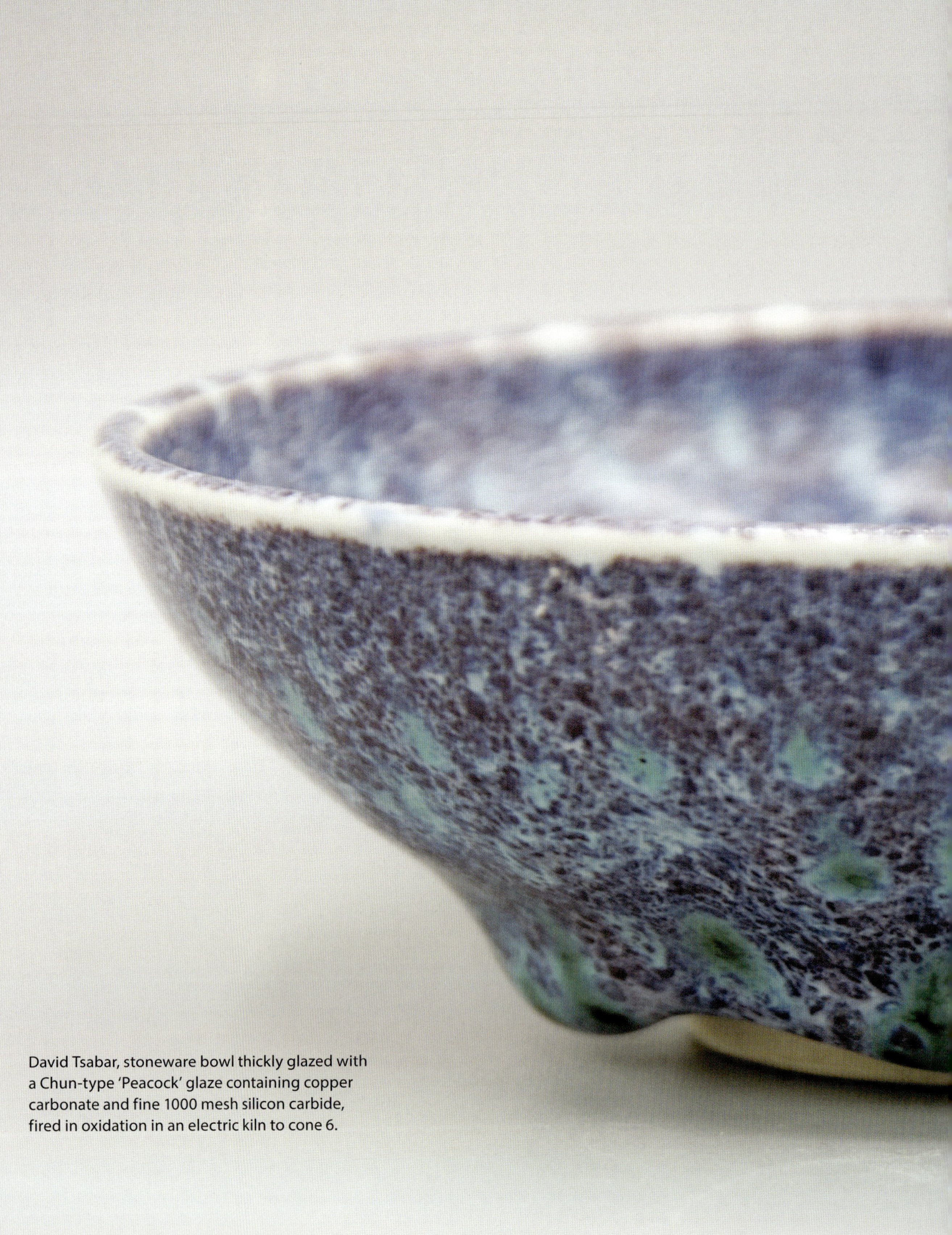

David Tsabar, stoneware bowl thickly glazed with a Chun-type 'Peacock' glaze containing copper carbonate and fine 1000 mesh silicon carbide, fired in oxidation in an electric kiln to cone 6.

SECTION 1
Glaze principles and application

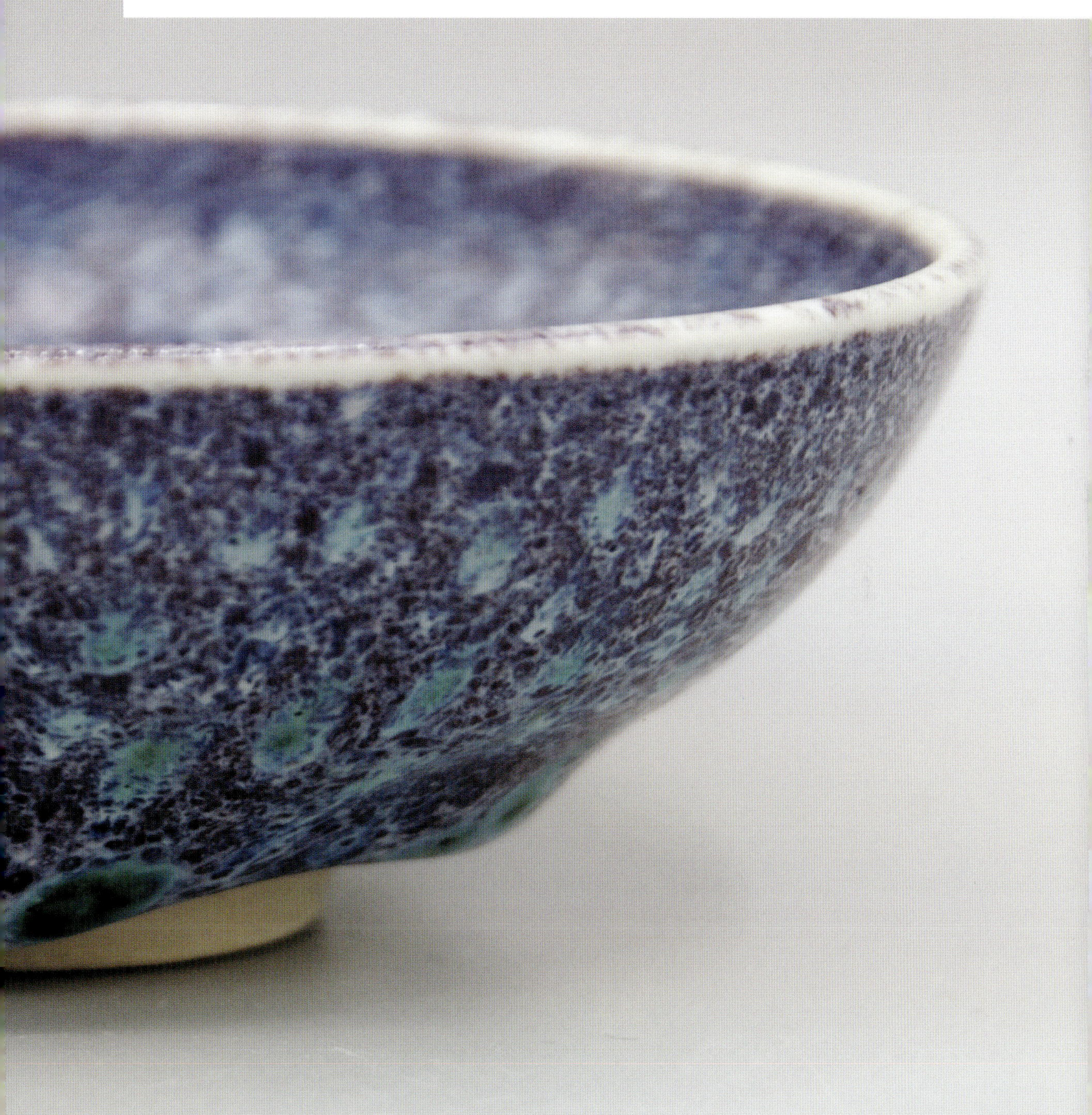

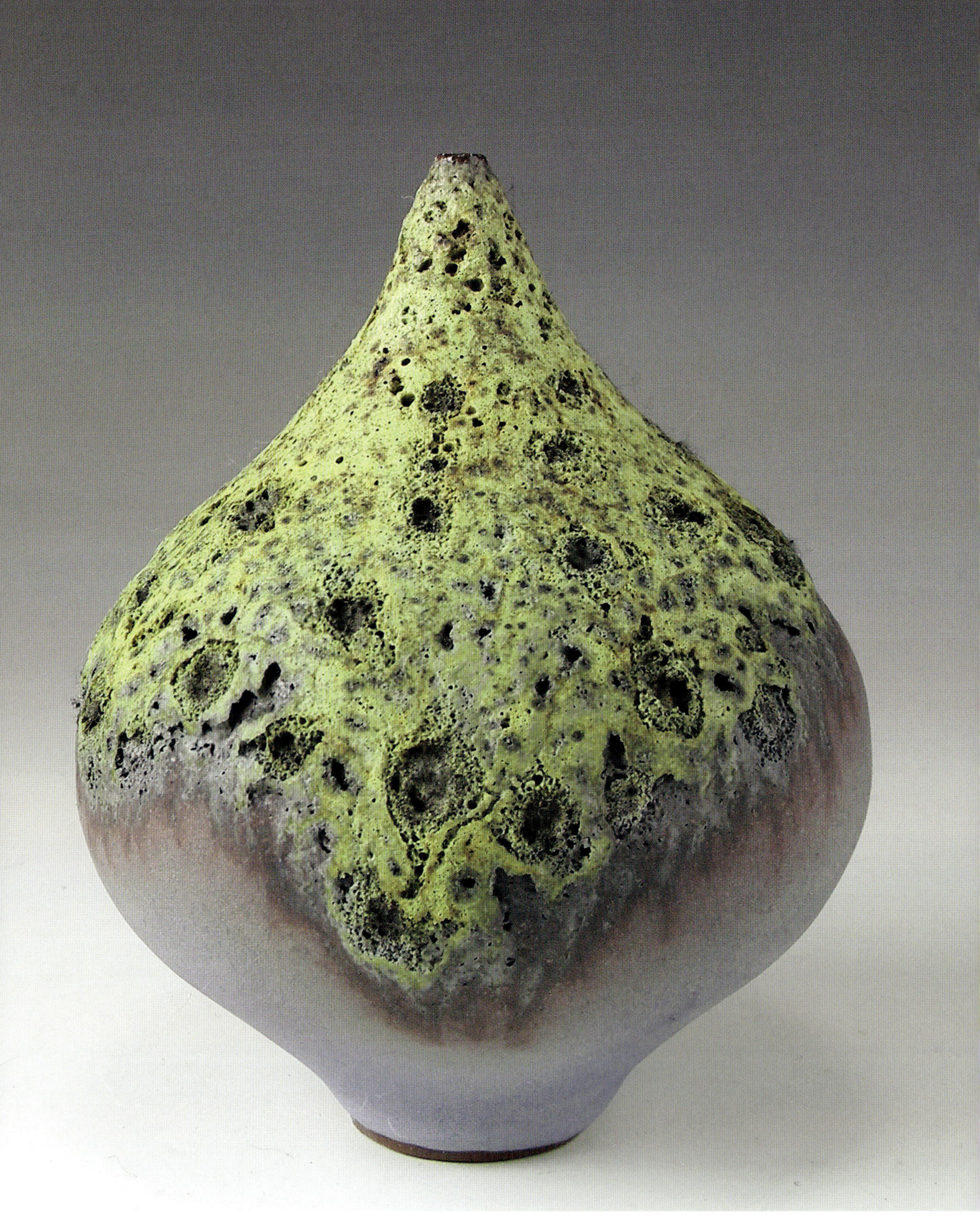

1

Understanding glazes

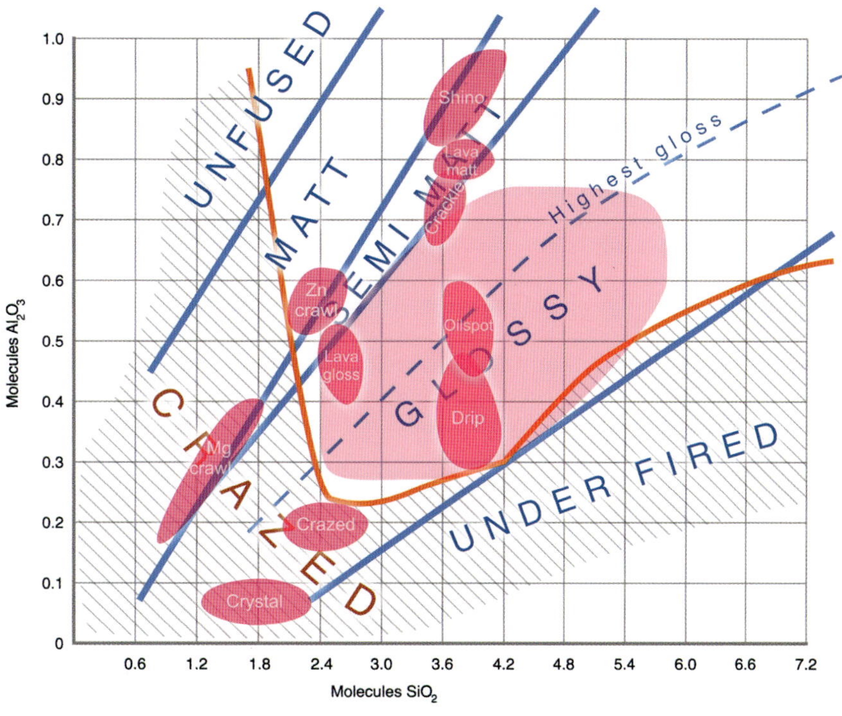

Diagram showing the alumina and silica range of various special-effect glazes. The effect of changing the alumina and silica in porcelain glazes fired to cone 11 with constant flux 0.3 K_2O and 0.7 CaO. The ratio of 1:5 alumina to silica gives a semi-matt glaze, while 1:8 gives a shiny glaze. The straight lines on the chart represent alumina:silica ratios of 1:4 (matt), 1:5 (semi-matt) and 1:12 (shiny, crazed glaze). The dashed line is 1:8 Al_2O_3:SiO_2 (bright, shiny glaze). The hatched area shows crazed glazes. The pale pink area shows Cooper and Royle's limits for stable glazes from cones 5 to 8. The special effects shown may move to the left for lower temperatures or to the right for higher temperatures. Data from R.T. Stull 1912. Graphics by Henry Bloomfield.

Recipe books can be very useful but if, owing to numerous variables, the glaze doesn't turn out how you expected, you need to understand the underlying principles in order to correct it. The many variables affecting glazes include clay body colour and texture, glaze materials used, glaze specific gravity, glaze application thickness, kiln type and size, firing temperature, firing time and atmosphere inside the kiln. All these factors contribute to the final appearance of the glaze.

This book is arranged in two sections: glaze principles including understanding and using glazes, and special-effect glazes including glaze recipes and explanations of how the special effects are created.

Special-effect glazes often lie outside the limits of what makes a glossy, transparent, functional glaze. On the diagram shown, the horizontal axis represents the number of molecules of silica in the glaze and the vertical axis shows alumina, found in clay and added to glaze to increase viscosity. The pale pink area shows the relative amounts of silica and alumina needed to make glossy glazes (together with the appropriate amount of fluxes needed to melt them at 1200–1260°C/2192–2300°F). Silica is the main glass former in a glaze, while alumina increases the viscosity and prevents the glaze from

Mike Hamlin, *Raindrop Vase*, 2017. Wheel-thrown red earthenware, satin matt and crater glazes, fired in electric kiln to 1184°C (2163°F) with a slow cooling cycle, 10 x 11.5 x 10cm (4 x 4.5 x 4in.).

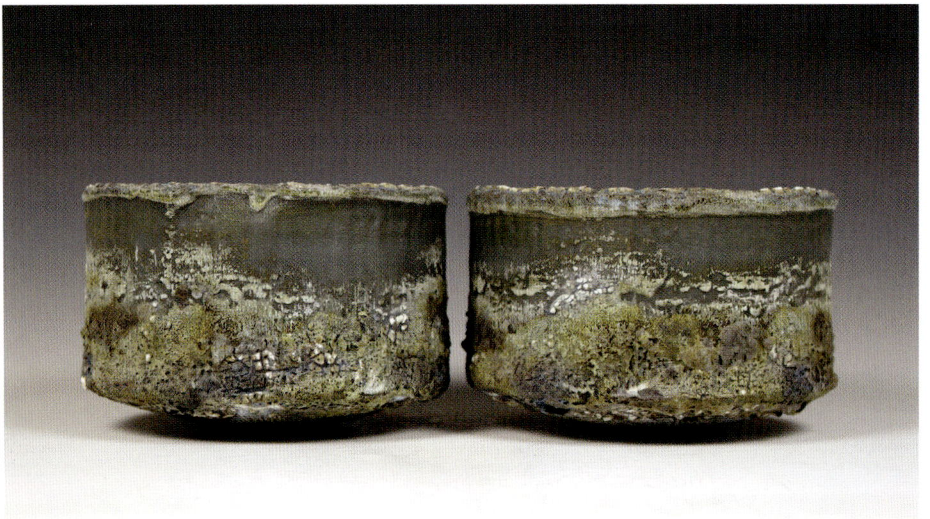

running off the pot. The proportion of silica to alumina needs to be balanced: if too much alumina is added, the glaze will become matt; too much silica and the glaze will not melt. Many of the various types of special-effect glaze lie outside the boundaries, such as crawl glazes, which can be cracked like a dry river bed or melted into beads. Volcanic glazes also lie in or near the semi-matt area; they need to be viscous enough to trap the bubbles of carbon dioxide gas coming from the silicon carbide added to the glaze. Shino glazes are often crawled as they are very high in clay, which shrinks on drying and forms cracks. Other special-effect glazes lie within the glossy area, such as oil-spot and drippy Chun glazes. These glazes are relatively high in silica, which enables any bubbles to escape and heal over. When the amount of flux in the glaze is high relative to the silica and alumina, the glaze becomes very runny and crystals can form during cooling, such as in crystalline glazes. Glazes with low silica are often crazed as they have a higher expansion than the clay body, which causes a network of fine cracks to appear upon cooling.

Many of these special-effect glazes are only for use on decorative or sculptural pieces, but oil-spot and drippy glazes can be used on functional tableware. Crawl glazes and lava glazes can be used on the outside of vases and bowls.

ABOVE: Paul Wearing, *Cylinders*, 2018. Handbuilt stoneware, brushed on slip and glazes containing magnesium carbonate, barium carbonate, vanadium pentoxide and silicon carbide, 10 x 15cm (4 x 6in.).

RIGHT: Katrina Pechal, *Dented Vase*, thrown stoneware, biscuit slip, volcanic glazes, height: 15cm (6 in.).

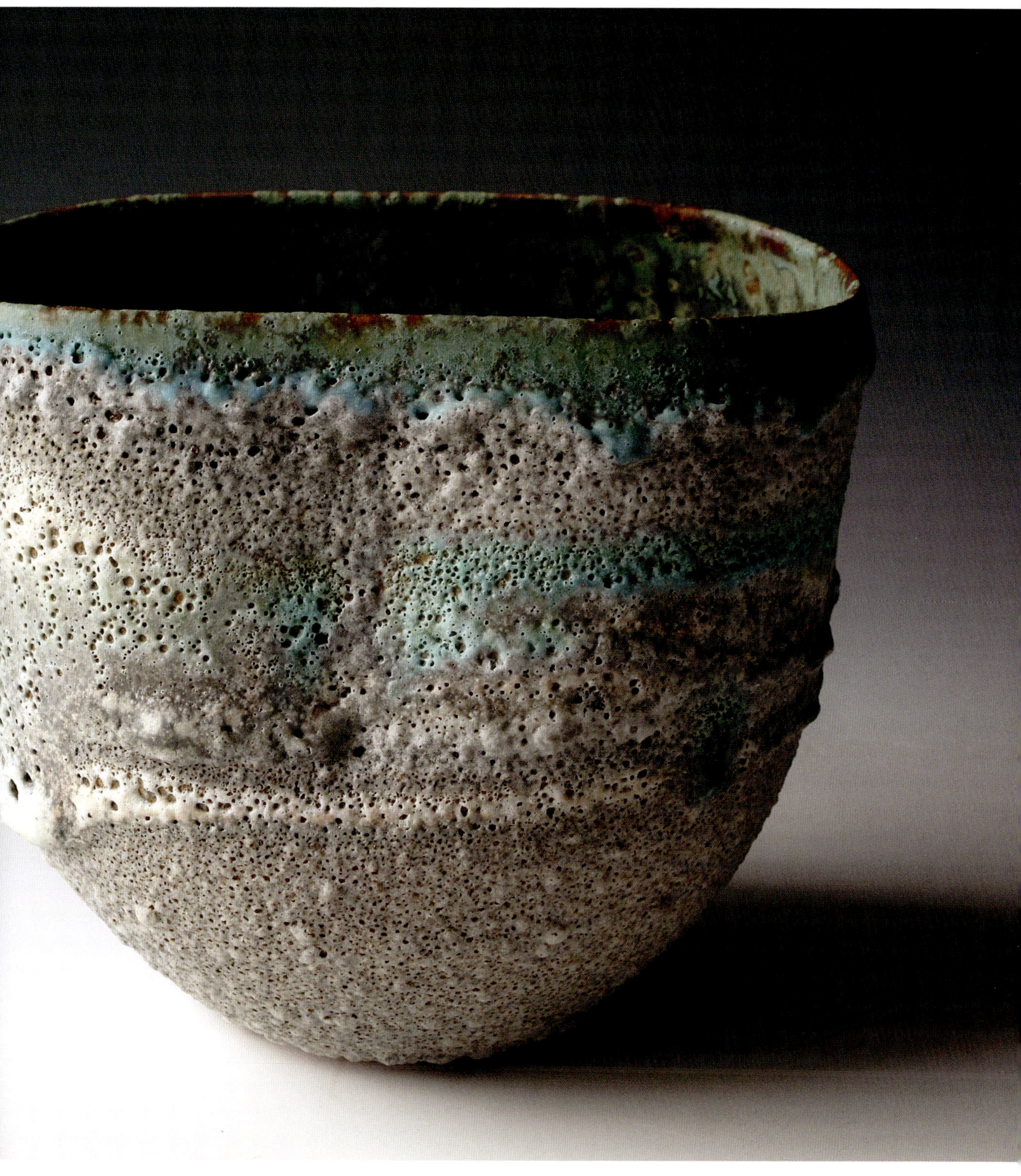

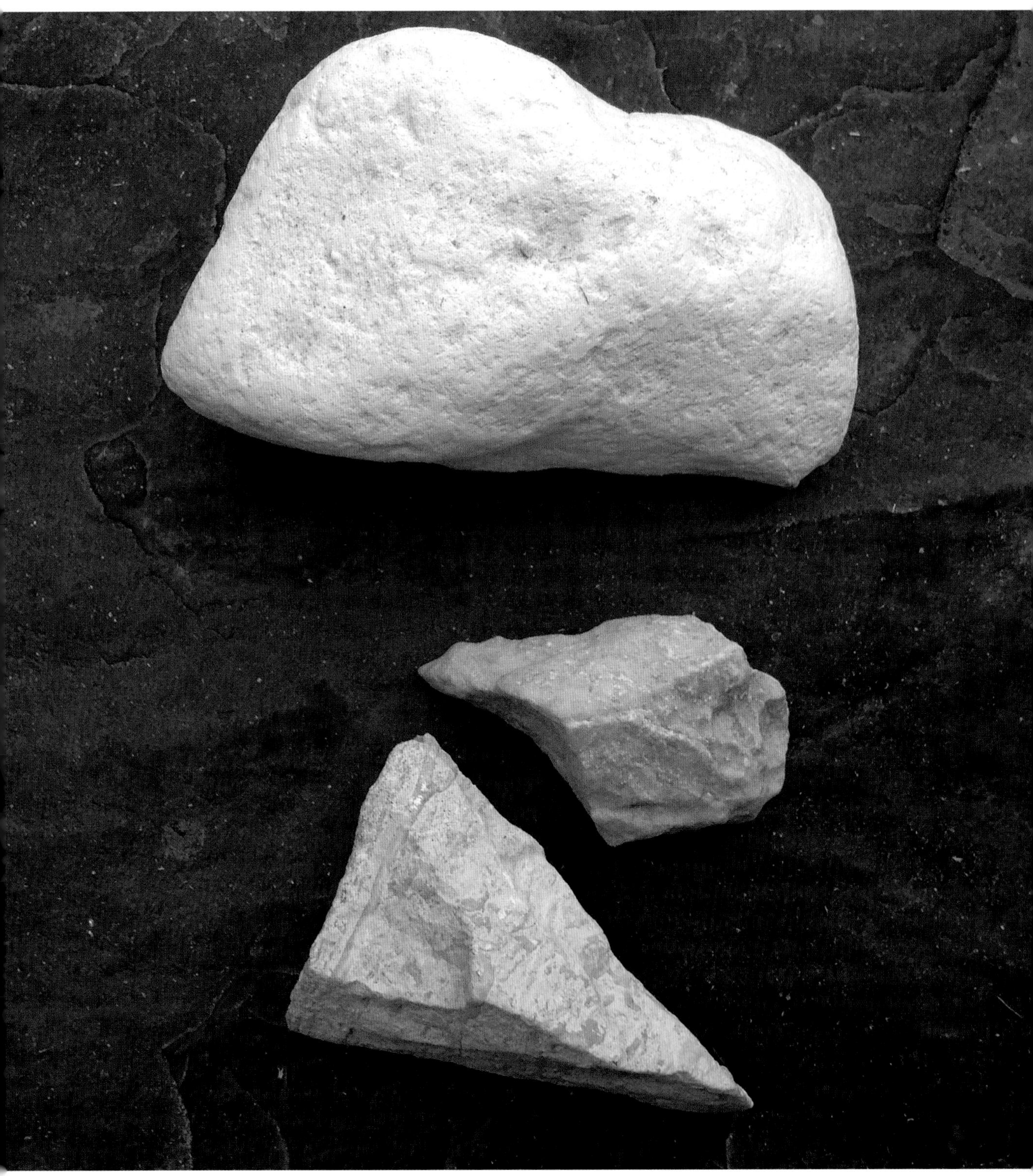

2

Glaze materials and minerals

The main material used in glazes is silica, in the form of ground flint or quartz. This is pure silicon dioxide (also known as silica) and is a glass former. The melting point of silica is too high for it to melt in a kiln, so fluxes are added to reduce the melting temperature. The main flux in stoneware glazes is feldspar, which contains sodium and potassium oxides. These are alkali metal oxides, which react with the acidic silica and break down the structure, helping it to melt. However, a simple mixture of sodium and silica, sodium silicate, would make a glaze which was soluble in water and likely to wash off, so a second flux is added to make the fired glaze insoluble. This secondary flux is usually whiting (calcium carbonate), found in chalk and limestone. Other secondary fluxes include the alkaline earths magnesium, barium, strontium and zinc. They help to strengthen the glaze and make it more stable. However, the resulting glaze would be very runny, so clay is added to make it more viscous in the melt. Clay contains aluminium oxide (known as alumina) and silica. China clay, ball clay and bentonite can all be used as glaze materials and are supplied in powdered form. These also help to suspend the heavier ingredients in water in the glaze bucket, preventing them from settling.

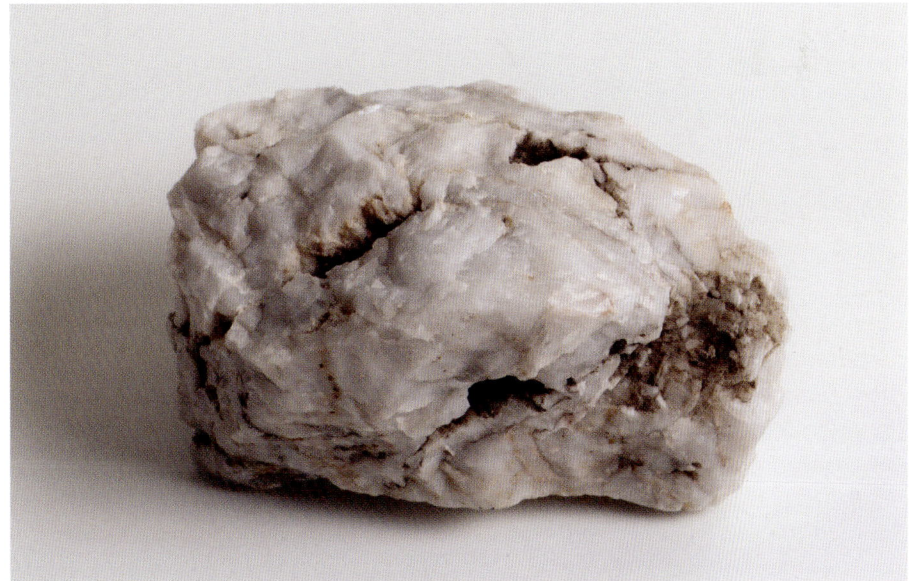

LEFT: Chalk from south-east England (at top) and limestone (smaller pieces) from the Alps.

RIGHT: Quartz from rock vein.

Silica (flint, quartz)	Glass former	}	Essential glaze materials
Flux (feldspar, whiting)	Melter		
Alumina (clay)	Stiffener		

Stoneware glaze recipe,
transparent glossy (1280°C/2336°F)
Potash feldspar 27
Whiting 21
Quartz 32
China clay 20

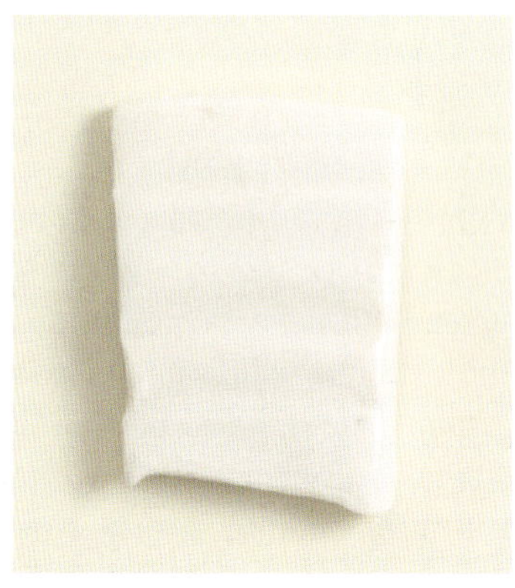

Basic glaze recipes; transparent glossy.

Earthenware glaze recipe,
transparent glossy (1100°C/2012°F)
Calcium borate frit 39
Soda feldspar 27
Whiting 5
Quartz 23
China clay 6

In the earthenware glaze recipe above, some frit has been added. This is a kind of man-made feldspar, containing fluxes such as sodium, calcium and boron oxides, as well as silica. The reason for making frits is that some glaze materials such as borax are water-soluble, eventually forming crystals in the glaze bucket that will not dissolve back into the glaze. A frit combines soluble materials with silica, which prevents them from dissolving in water. The more frit that is added to a glaze, the lower the firing temperature will be. This is because frits contain powerful fluxes such as sodium and boron, the latter of which acts as a low-temperature glass former. Mid-range glazes (between stoneware and earthenware firing temperatures) can contain between 10–30% frit, while earthenware and raku glazes can contain up to 90% frit.

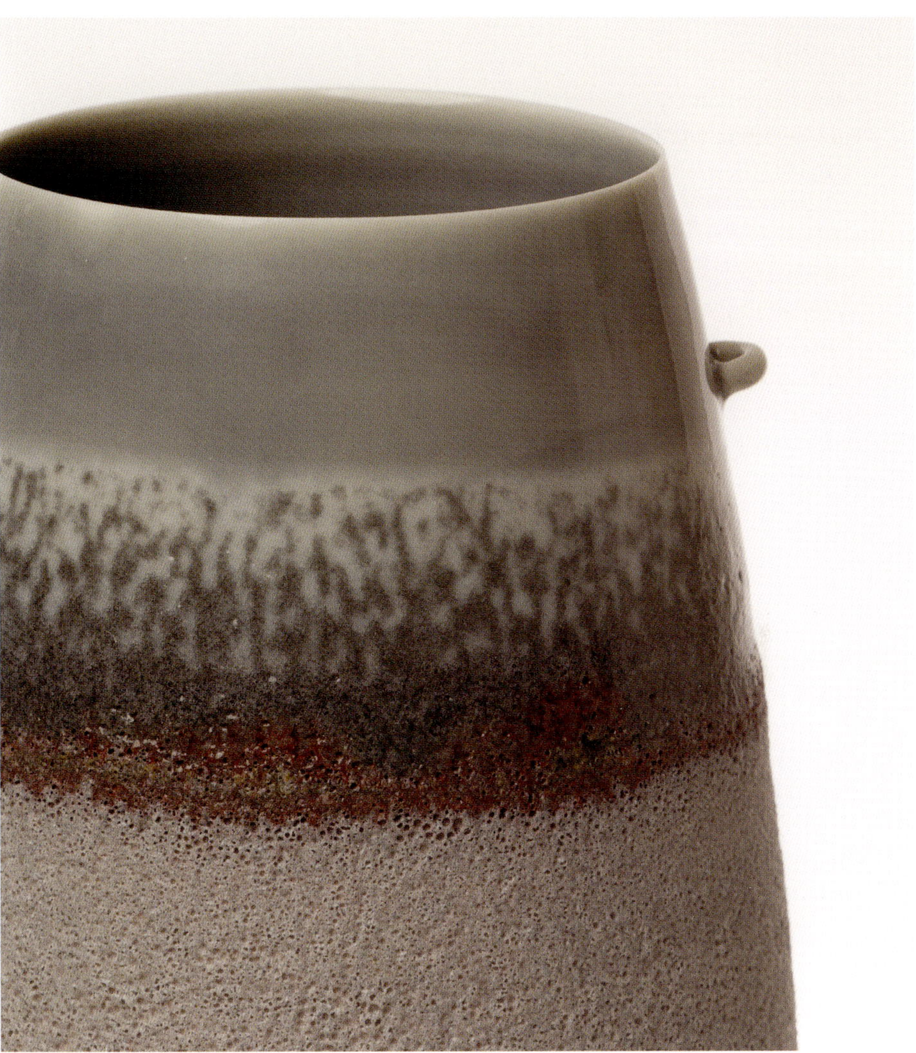

Jacqui Ramrayka, Seascape vessel, thrown porcelain, volcanic dolomite matt glaze under glossy grey glaze, fired to 1260°C (2300°F). *Photo: Dee Honeybun.*

Glaze Properties

To make a shiny glaze, the glaze materials are combined together in a eutectic mixture. This is the combination of materials mixed in proportions such that this mixture has the lowest melting temperature. Because this is the combination most likely to be fully melted, it will achieve the most glossy glaze from the materials. For silica and alumina, the eutectic ratio is 9 molecules silica to 1 molecule alumina; this mixture melts at a lower temperature than either of the pure materials alone. Alumina is found in clay, together with silica (a clay molecule is made of one alumina to two silica and two water molecules). Alumina is also present in feldspar (one alumina to six silica and one soda or potassia molecule). Glazes with a ratio of 1:9 alumina to silica (plus enough fluxes to melt them) will be shiny and transparent. Reducing the silica to a molecular ratio of 1:5 alumina to silica will make a matt glaze.

Altering the glaze

To make the glaze melt at a lower temperature, a frit can be added. Many frits contain fluxes and boron, a glass former that melts at a lower temperature than silica.

To make the glaze more shiny, the clay in the recipe can be reduced. This will reduce the viscosity of the glaze and make it more runny. For a glossy glaze, the optimal ratio of alumina to silica is around 1:8.

To make the glaze more matt, the silica can be reduced or more clay can be added until the ratio of alumina to silica is 1:5. This is known as an alumina matt or 'true matt' as it will remain matt even if fired to a higher temperature, as opposed to being matt because it is under fired.

Another way to make a matt glaze is to add more calcium or magnesium in the form of dolomite. These are known as lime matt or magnesium matt glazes. However, these matt glazes will have subdued colour.

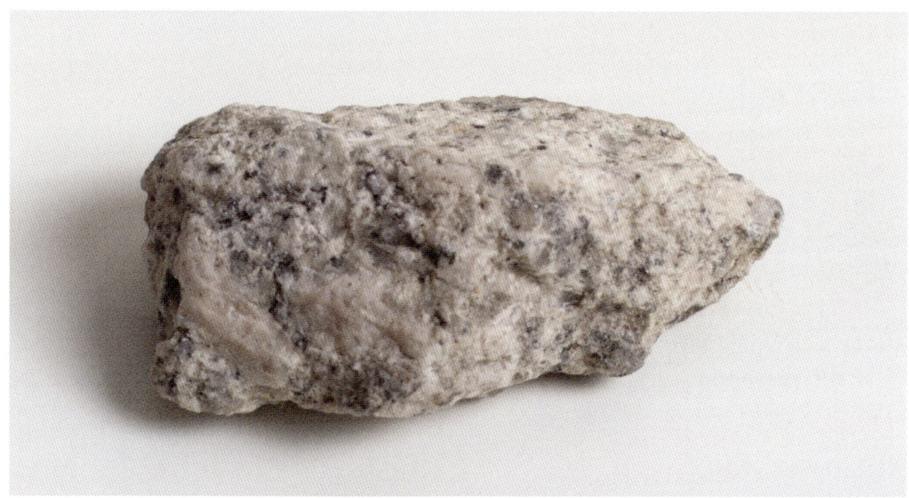

Granite from Cornwall, containing potash feldspar (pink or white crystals).

To make brightly coloured matt glazes, add barium or strontium carbonate. These will give bright turquoise with copper oxide or lime green with chromium oxide. These bright colours occur in most 'alkaline' glazes, i.e. those containing sodium, lithium, barium or strontium.

The alkaline earth (Ca, Mg, Ba and Sr – see p.154) matt glazes are generally known as flux matts but may become glossy if fired to higher temperatures unless they are also high in alumina (with the ratio 1:5 alumina to silica).

Looking at the glaze recipes on p.18, the differences between stoneware and earthenware glaze recipes can clearly be seen. The earthenware glaze has less clay, quartz and whiting. Potash feldspar has been replaced by soda feldspar and borate frit has been added to decrease the firing temperature. Soda feldspar gives a more fluid melt than potash feldspar, but the glaze will be slightly softer and will scratch more easily. In general, the higher the firing temperature, the harder and more scratch-resistant the glaze will be and the harder and stronger the fired clay body becomes.

Glaze materials

Feldspar

Most stoneware glazes are made using feldspar as the main flux. Feldspar contains sodium and potassium oxides. The feldspar is named for whichever is the dominant oxide: potash feldspar or soda feldspar. Cornish stone is a type of potash feldspar which is particularly high in silica. Nepheline syenite is a mineral related to feldspar and is particularly high in sodium (with only two silica to one alumina and one soda molecule, compared to six silica in feldspar). Some feldspars such as petalite and spodumene contain lithium, a very active flux, which is also available in its more concentrated form as lithium carbonate. Feldspar compositions may vary depending on where they are sourced from.

Silica

Silica is silicon dioxide, supplied to potters as flint or quartz. Either material can be used in glazes. Silica is also present in feldspar, clay, talc, wollastonite and frits. Quartz and flint are sold with a small amount of water added to damp down the dust. A respirator mask should be worn when handling the dry materials.

Clay

Clay is added to the glaze in its powdered form. Either china clay, ball clay or bentonite can be used. China clay is used in porcelain glazes such as celadons, where pure colours are important, while ball clay can be included in glazes for use on stoneware clays. Bentonite, a very plastic clay, is added in small amounts (2–3%) to glazes which are particularly low in clay. Clay helps to suspend the other ingredients in water, preventing them from settling. It also improves the raw strength of the glaze and prevents it drying to a powder on the pot. However, excess clay in a glaze can cause crawling (see Chapter 9).

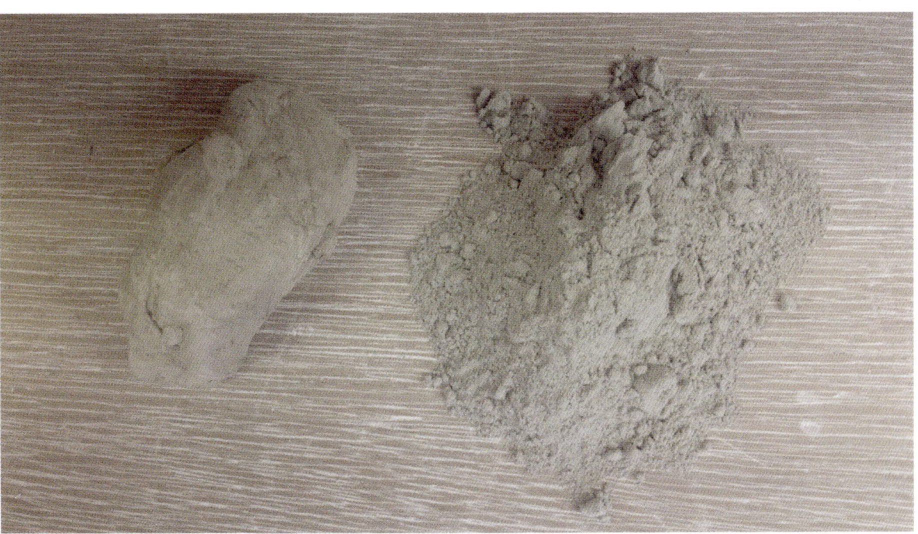

Clay and powdered bentonite.

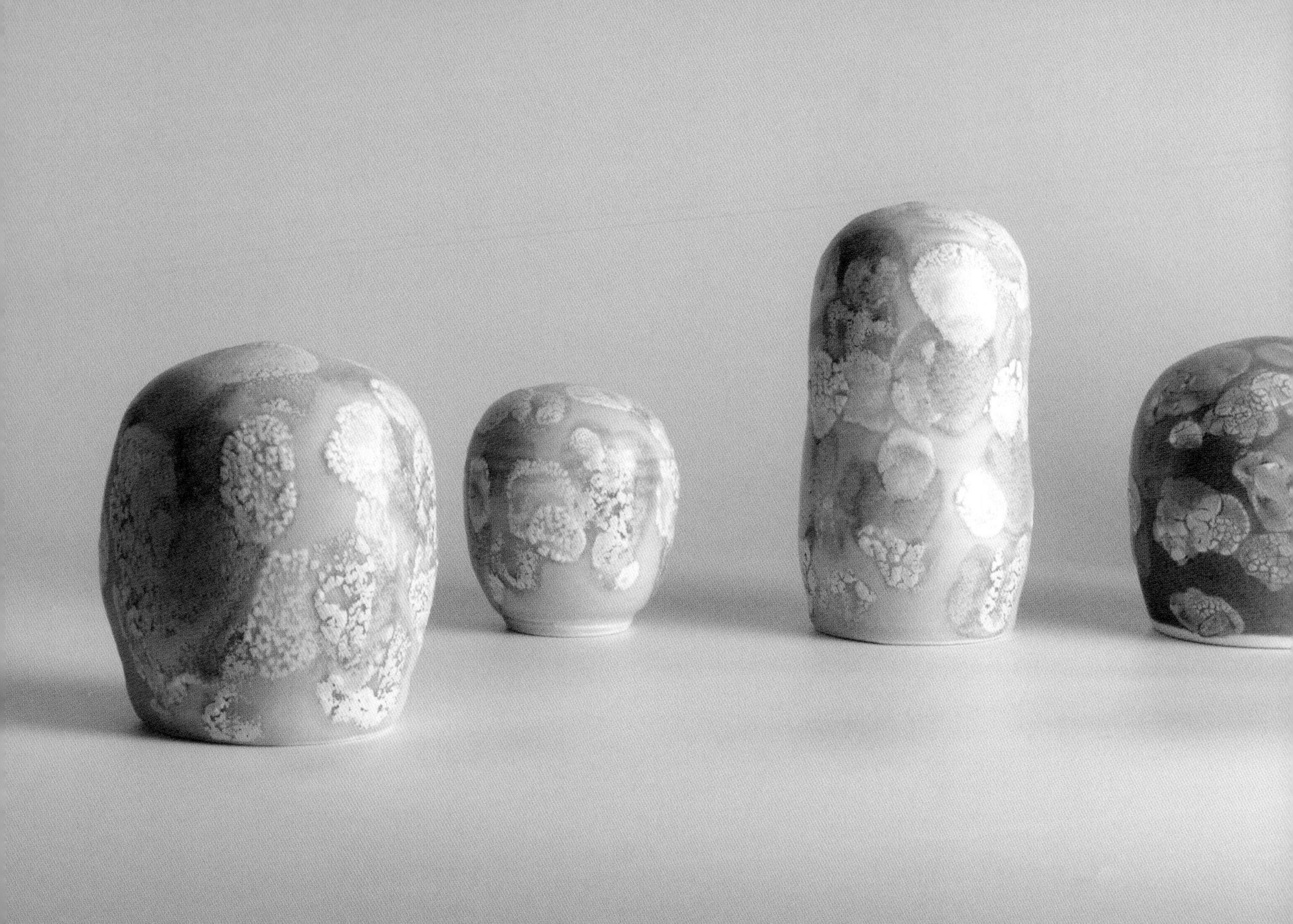

Calcium carbonate

The most common secondary flux is whiting, another name for calcium carbonate or chalk. Calcium can also be suppled by wollastonite (calcium silicate) and dolomite, (calcium magnesium carbonate). Excess calcia can cause the glaze to become matt. However, calcium matt glazes will generally have pale, bleached colours.

Magnesium carbonate

Magnesium is found in dolomite and talc. Dolomite is magnesium calcium carbonate and is added to glazes to make them satin matt. Talc is magnesium silicate, which has a similar matting effect, while also providing silica to the glaze, and is helpful in correcting crazing. Light magnesium carbonate is added to special-effect glazes to cause them to shrink and crawl.

Barium and strontium carbonate

Barium carbonate and strontium carbonate are added to glazes to make brightly coloured matts and volcanic glazes. The colours include turquoise from copper oxide and chartreuse green from chromium oxide. Barium carbonate is toxic but strontium carbonate is safer to use.

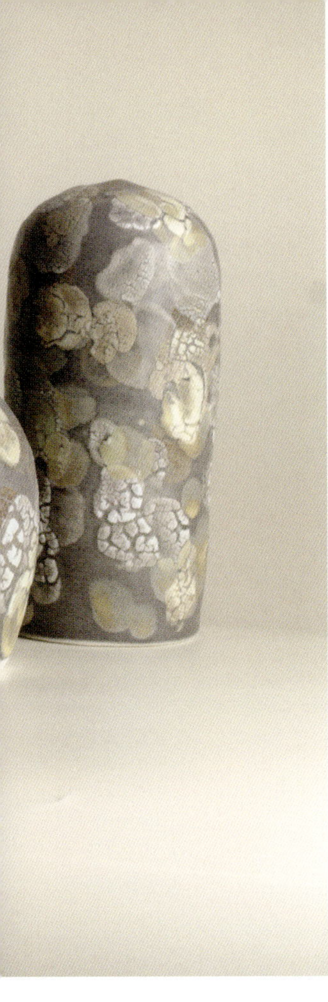

Zinc oxide

Zinc oxide is a flux used to make crystalline glazes. It can also be used to encourage crawling, and can form melted beads of crawled glaze.

Bone ash

Bone ash is calcium phosphate, and can be used to replace whiting in glazes in small quantities. Phosphorus causes opacity and mottling in glazes such as Chun glazes.

Wood ash

Wood ash is a source of calcium, magnesium, phosphorus and iron. When mixed with water, it is highly caustic, so rubber gloves should be worn. It can be used instead of whiting in a glaze but may need to be washed and sieved before doing so. See also Ash glazes chapter, p.89.

Frit

Frits are used as low to mid-temperature fluxes when the materials would otherwise be water-soluble and cause problems in the glaze. Materials that are made into frits include lead oxide, sodium carbonate and borax (sodium borate). These materials are mixed with silica and melted, and the resulting glass is cooled and ground to a powder. As they have already been fired, they melt more readily than unfritted materials. Owing to toxicity, lead frits are now less commonly used and borate frits have become more widely used. Frits are useful for lowering the melting temperature of cone 10 (1300°C/2372°F) glazes down to cone 8 or 6 (1260–1240°C/2300–2264°F).

ABOVE: Linda Bloomfield, porcelain with magnesium matt glazes and lichen-effect glazes using magnesium carbonate.

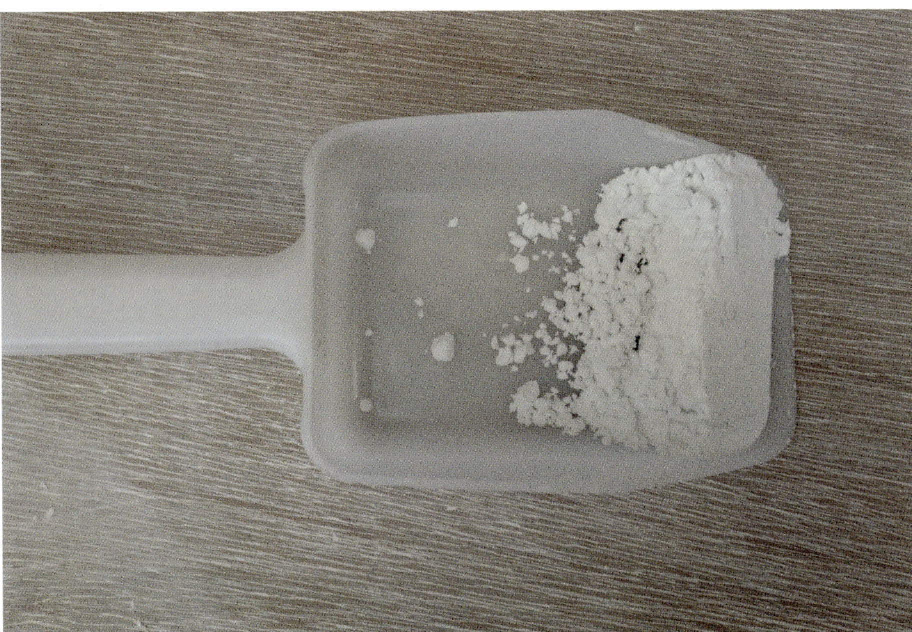

Borax frit (containing sodium, calcium borate and silica).

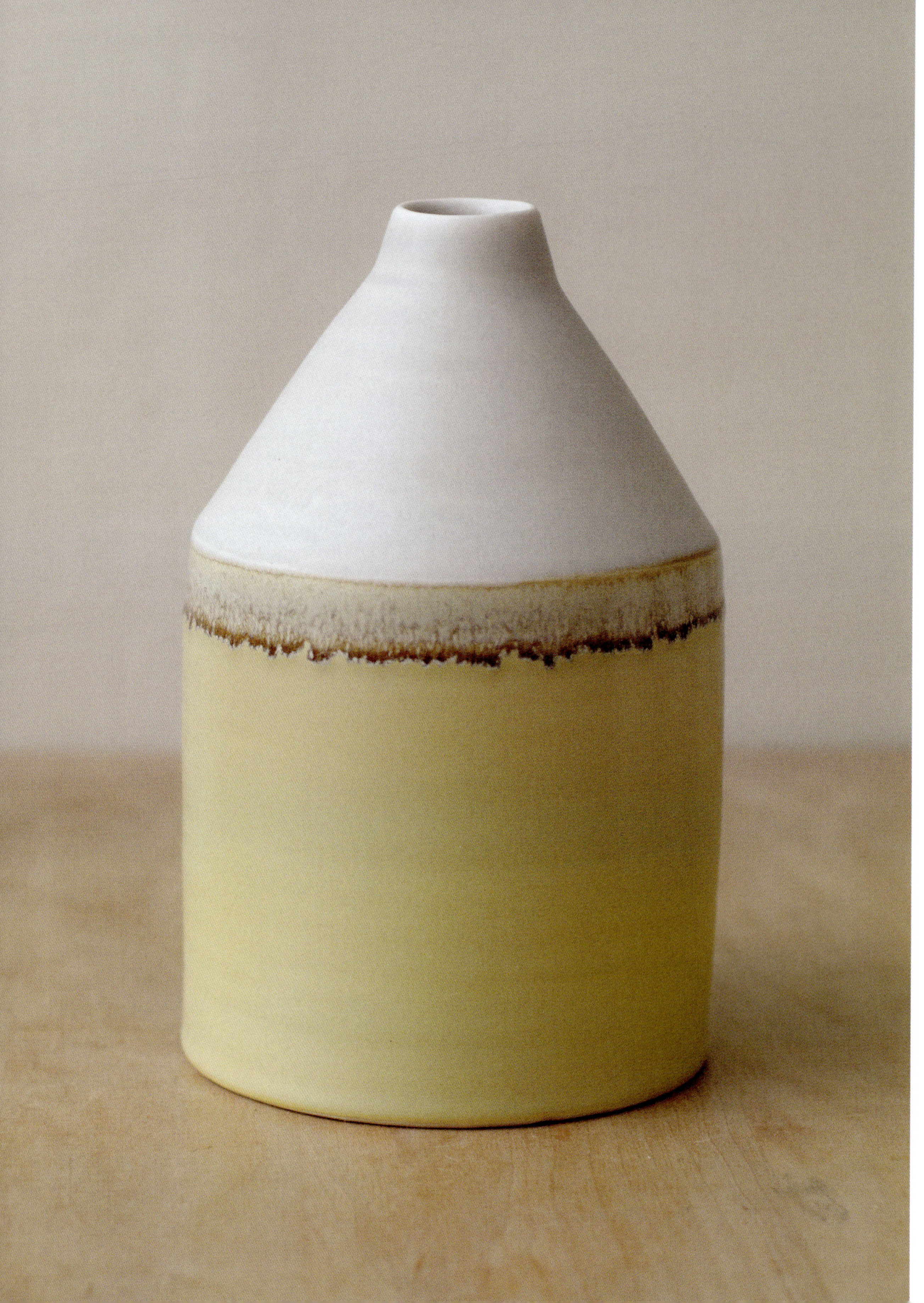

3

Colouring glazes

Glazes can be coloured by adding commercial stains or colouring oxides. Stains and underglaze colours have been mixed with stabilisers and opacifiers and have already been fired, so their colour will not change further during firing. However, colouring oxides are likely to change colour during firing as they dissolve and react with the materials in the glaze, giving transparency and depth. The colouring oxides include cobalt, copper, chromium, iron, manganese, nickel, rutile and vanadium. Oxides can be brushed directly on to biscuit-fired ware and covered with a clear glaze, or mixed in with the glaze to give more uniform colour. When they are added to a base glaze recipe, they are added as an extra percentage amount. For example, if a recipe has 2% copper oxide, you will need to add 2g to 100g dry weight of base glaze. Once mixed with water, the glaze will need to be sieved several times using an 80 to 100 mesh sieve (meaning it has 100 holes per linear inch). If speckles still remain, you may need to sieve again through a 120 mesh sieve. Many of the colouring oxides are toxic, so a mask must be worn when weighing the dry ingredients and firing fumes should be avoided.

The fluxes in the base glaze will have an effect on the colour developed during firing. For example, magnesium-based glazes will make green with copper oxide, while barium-based glazes will be turquoise or blue. The kiln atmosphere will also affect the colour: restricting the oxygen intake to create a reduction firing will cause copper oxide to turn oxblood red and iron oxide to turn celadon blue-green. Dolomite matt glazes tend to have subdued colour, while barium matts can have bright colours including turquoise from copper, lime green from chromium, and purple from nickel oxide. Barium carbonate is toxic, so it may be preferable to substitute with non-toxic strontium carbonate. When substituting strontium for barium, use approximately 75% of the amount, and always test first. A small addition of lithium carbonate can also brighten the colours obtained from oxides.

In general, small amounts of colouring oxide will dissolve in the glaze, resulting in a transparent coloured glaze. Larger amounts will remain as undissolved particles and the glaze will be opaque. In matt glazes, the matt surface is often made of tiny crystals, which will be coloured by oxides. Crystals can be encouraged by adding rutile or titanium dioxide, which remain unmelted and act as seeds for crystal growth.

LEFT: Linda Bloomfield, Two-tone bottle, thrown porcelain, mustard and grey matt glazes, height: 23cm (9in.), fired to 1250°C (2282°F).

Cobalt

Cobalt oxide is named from the German kobold; goblins once thought to live in mines that produced toxic ores (cobalt is pronounced to rhyme with low vault). Cobalt is toxic and it is found in minerals such as smaltite in combination with arsenic. In glazes, cobalt oxide or carbonate give a strong blue colour, which tends towards purple except in zinc-based glazes, where it is a cool, bright blue. The carbonate is a weaker form and 1.5x the amount of oxide is needed to give the same strength of colour. Amounts of 0.5–2% cobalt oxide will give a strong blue in glazes, which can be subdued by adding iron and manganese oxides. Black glazes can be made using 2% each of cobalt, manganese, iron and 1% nickel oxide. Matt glazes containing dolomite or talc will turn cobalt a lavender blue colour.

Copper

Copper oxide will give green in magnesium-based glazes and turquoise in calcium or barium glazes fired in oxidation, or oxblood red when fired in reduction. Copper carbonate is a weaker form and can be substituted for copper oxide but 1.5x is needed. In calcium, barium or strontium glazes, 0.5–1% copper oxide will give a turquoise colour (or pale blue when combined with tin oxide), 2–3% will be more green but more than 4% will produce a black, metallic effect. Copper oxide CuO becomes volatile at 1025°C (1877°F) in the kiln and can cause a pitted texture in the glaze as it releases oxygen and reduces to cuprous oxide Cu_2O. When covered with a white opaque glaze, this can cause a spotted effect.

BELOW, LEFT TO RIGHT:

Cobalt oxide in ash glaze and Chinese porcelain and cobalt arsenide mineral skutterudite. Blue ash glaze; potash feldspar 40, wood ash 60, bentonite 2, cobalt carbonate 0.1 (fired to cone 9).

Copper oxide in potash and soda glazes and copper silicate mineral chrysocolla. Glaze recipes: Turquoise glaze: soda feldspar 47, quartz 18, calcium borate frit 15, whiting 14, china clay 5, copper oxide 1 (fired to cone 8). Green ash glaze: potash feldspar 40, wood ash 60, copper oxide 1 (fired to cone 9).

Chromium oxide and chrome diopside silicate mineral including quartz crystals. Chrome green glaze: soda feldspar 47, quartz 18, calcium borate frit 15, whiting 14, china clay 5, chromium oxide 0.5 (fired to cone 8).

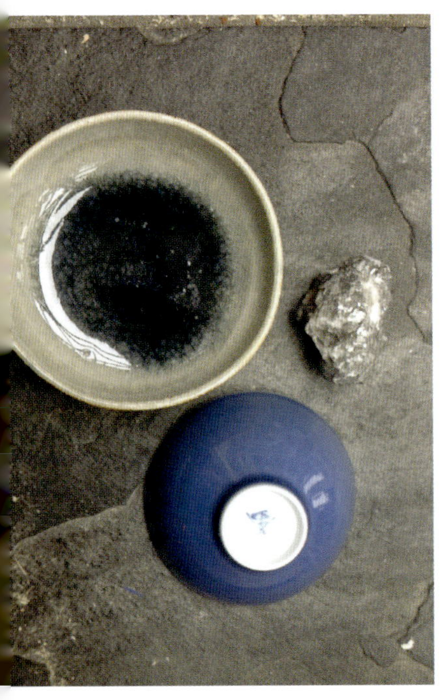

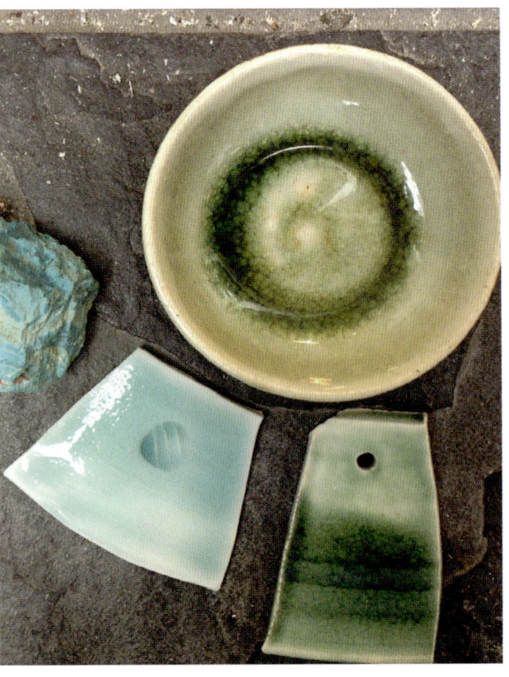

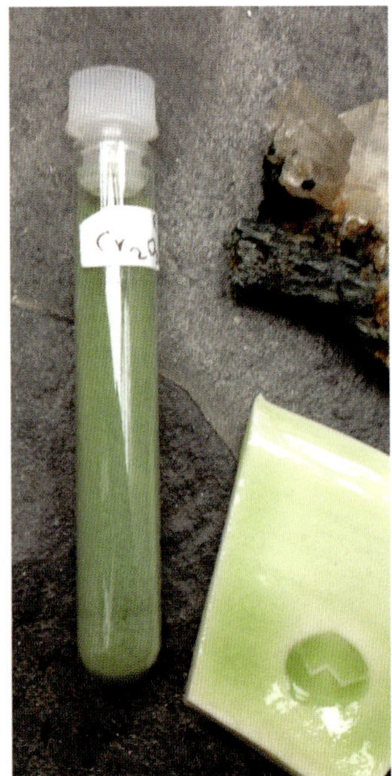

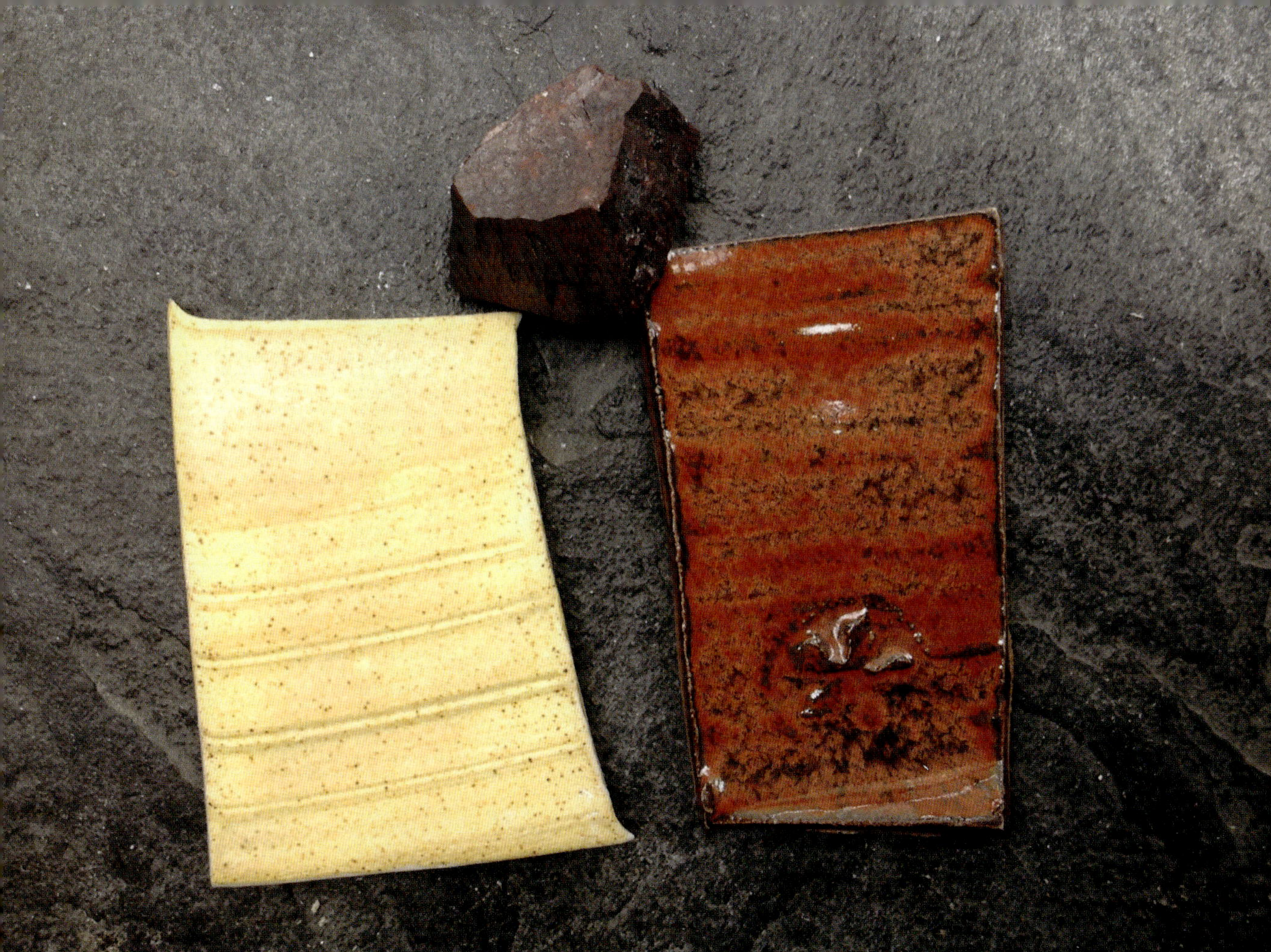

ABOVE: Iron oxide in red and yellow glazes and hematite mineral. Yellow matt glaze: Nepheline syenite 42.5, dolomite 15.5, barium carbonate 24, china clay 9, quartz 9, zirconium silicate 19, red iron oxide 3.5. Red iron glaze: Potash feldspar 47, talc 17, bone ash 15, quartz 11.5, china clay 6, lithium carbonate 4, red iron oxide 11.5 (fired to cone 8).

Chromium

Chromium oxide gives a range of colours in glazes. The usual colour from chromium is bright green, but a small amount (0.1–0.5%) will give a pink or red colour with tin oxide (5%), known as chrome-tin pink. Chromium tends to turn brown in the presence of zinc oxide, except when the glaze is high in zinc and alumina, where it will form a pale pink colour without adding tin oxide. Chromium oxide is toxic.

Iron

Iron oxide is used in many traditional glazes including honey yellow, celadon green-blue, khaki red-brown and tenmoku brown-black. 0.5–15% iron oxide can be added to glazes, larger amounts giving stronger colours. Black and grey glazes can be made using various combinations of iron, manganese, nickel and cobalt oxides. Iron oxide is present in rust and red earthenware clay. Iron oxide pigments used by potters include red iron oxide, black iron oxide and yellow iron oxide. Red iron oxide Fe_2O_3 decomposes to black iron oxide Fe_3O_4 during firing at 1232°C (2250°F), giving off bubbles of oxygen gas in high-iron glazes to form an oil-spot effect (see also Chapter 18).

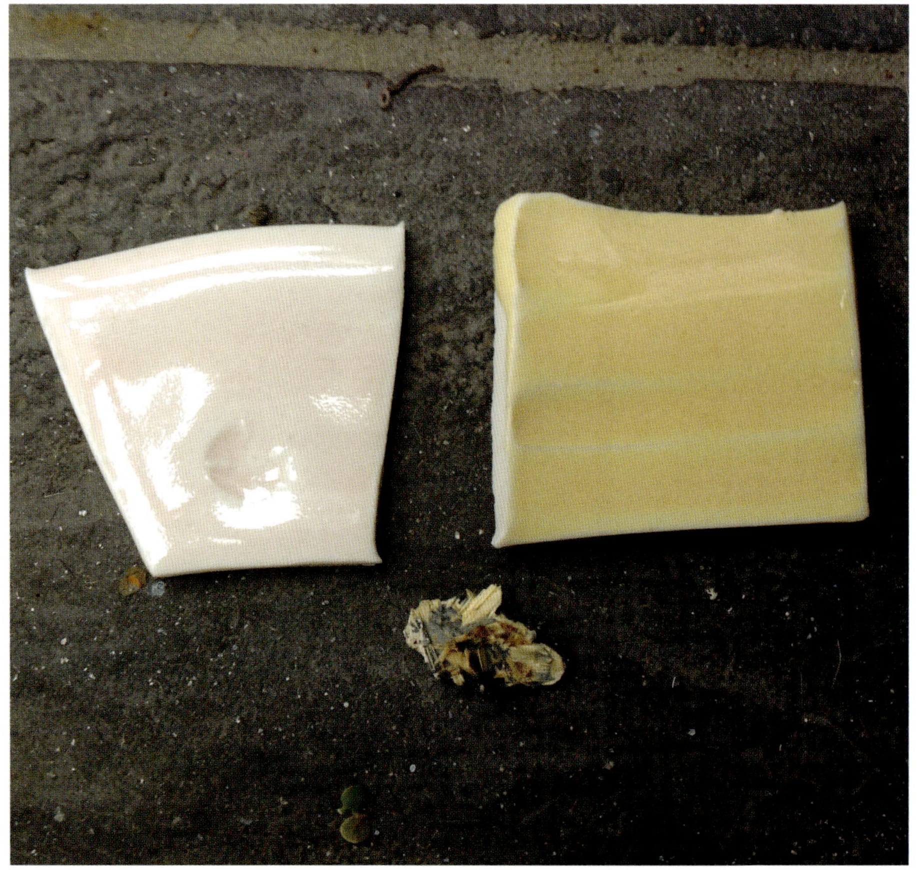

LEFT: Rutile-tin pink and yellow glazes and needle-shaped crystals of rutile. Pink glaze: soda feldspar 47, quartz 18, calcium borate frit 15, whiting 14, china clay 5, tin oxide 4, rutile 2. Yellow glaze: potash feldspar 33, talc 21, quartz 16, china clay 15, whiting 12, tin oxide 5, rutile 7 (both fired to cone 8).

OPPOSITE, TOP RIGHT: Vanadium pentoxide in matt glazes and vanadinite. Yellow matt glaze potash feldspar 50, dolomite 20, china clay 20, bone ash 10, vanadium pentoxide 5; for green add 5 zirconium silicate (fired to cone 8).

BOTTOM LEFT: Manganese dioxide in a barium glaze and silicate mineral rhodonite. Pink-brown glaze: FFF feldspar 37, barium carbonate 37, lithium carbonate 3, quartz 15, china clay 5, manganese dioxide 2 (fired to cone 8).

BOTTOM RIGHT: Nickel oxide in green and mustard satin matt glazes and nickel arsenate mineral annabergite. Nickel crystal glazed pot by Avril Farley. Green glaze: potash feldspar 33, talc 21, quartz 16, china clay 15, whiting 12, zinc oxide 3, nickel oxide 3, titanium dioxide 5. Yellow glaze: add titanium oxide 5 (fired to cone 8).

Rutile

Rutile is a mineral containing titanium dioxide and up to 10% iron oxide. A similar mineral containing 50:50 titanium and iron oxide is called ilmenite. 2–10% rutile can be used to give streaked and mottled effects in glazes; see the following chapter on rutile.

Nickel

Nickel oxide can be used, together with cobalt, to make grey glazes. It will produce green and mustard colours with titanium dioxide in magnesium glazes or pink and steel blue in barium and zinc glazes. Nickel oxide is toxic and only 0.1–3% is needed in a glaze.

Manganese

Manganese dioxide will produce a dark brown in quantities of 1–15%, giving metallic surfaces above 20%. It will give a pinkish brown in barium glazes and purple when combined with a small amount of cobalt. Manganese dioxide MnO_2 decomposes to manganese oxide MnO at 1080°C (1976°F), giving off gas and becoming a flux. The firing fumes are toxic.

Vanadium

Vanadium pentoxide can be used in magnesium, barium and strontium glazes to give a yellow or green colour. 2–8% is needed in the glaze. Vanadium pentoxide is slightly soluble and very toxic.

Rare earth oxides

The rare earths include cerium, praseodymium, neodymium, holmium and erbium. They are not strong colourants but can be used on porcelain to give pale yellow, green, lavender and pink. Holmium changes colour from yellow to pink under different lighting conditions, while neodymium changes from lavender in daylight to pale blue in fluorescent light.

BELOW, LEFT: Monazite sand and rare earth oxides neodymium, praseodymium and erbium 6% in transparent glossy glaze: soda feldspar 47, quartz 18, calcium borate frit 15, whiting 14, china clay 5 (fired to cone 8).

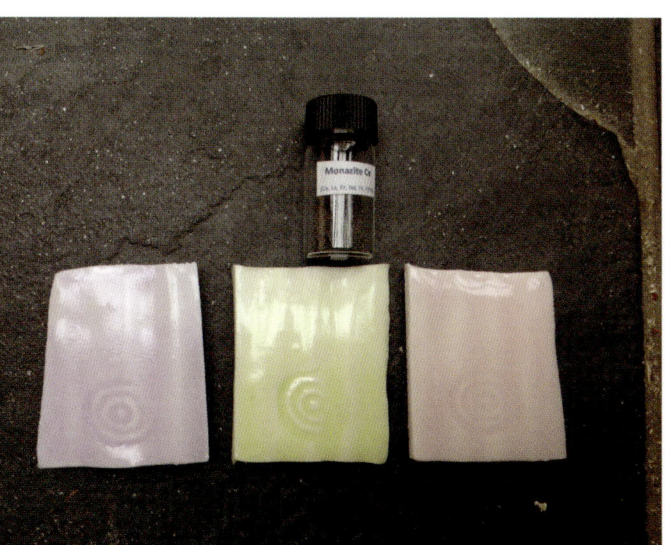

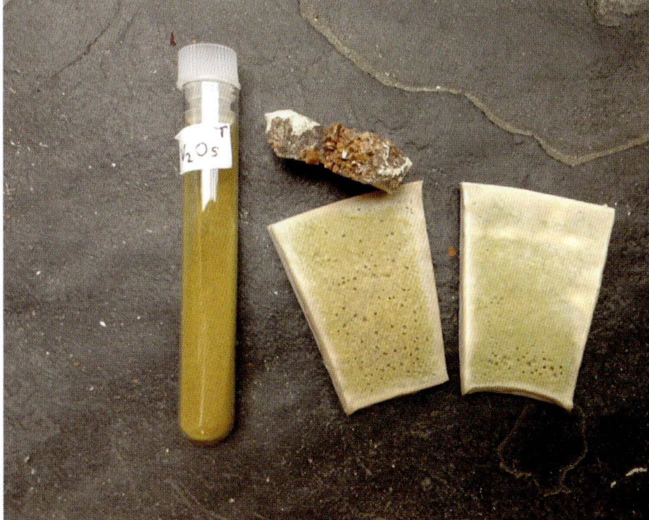

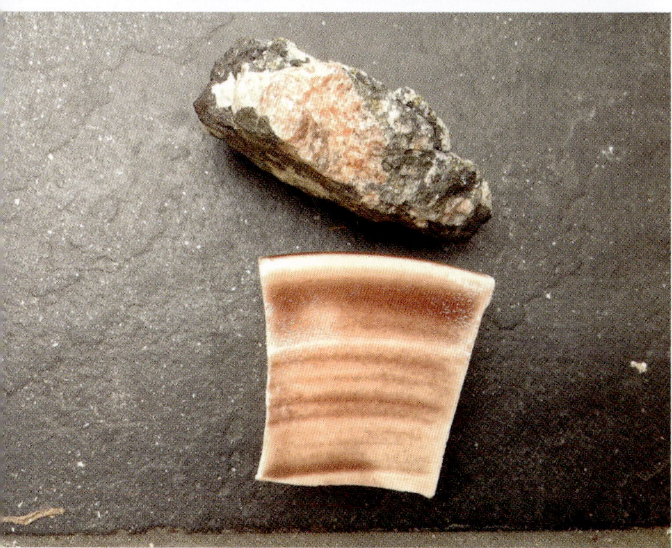

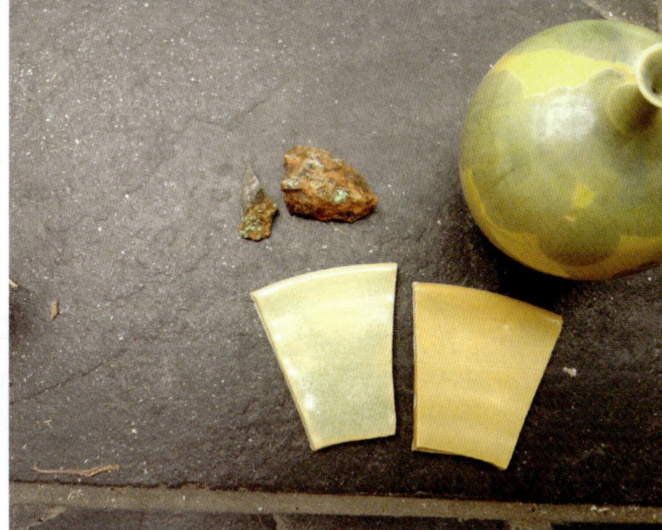

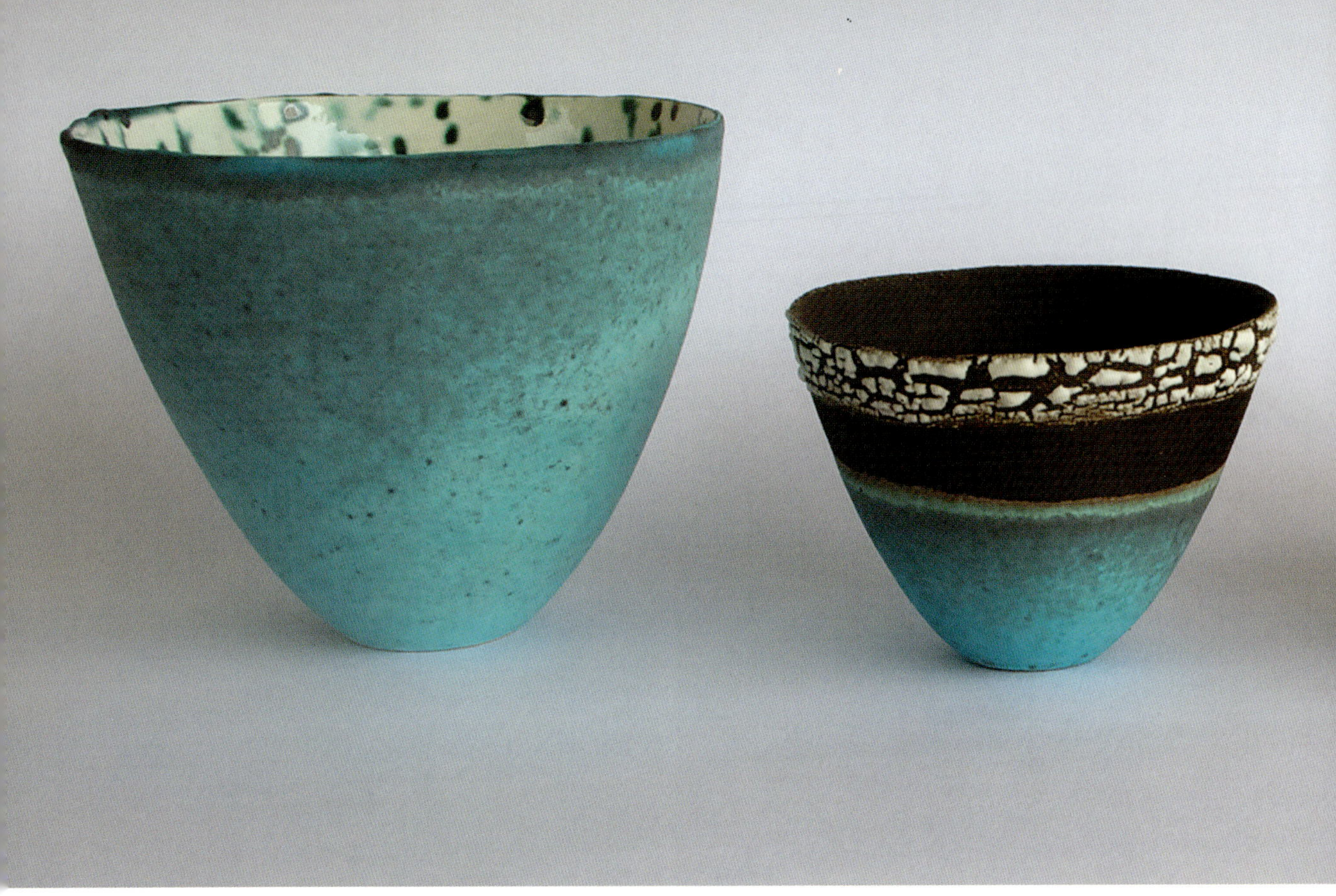

Opacifiers

White glazes are made by adding tin oxide or zirconium silicate to a clear glaze. Amounts of 5% tin oxide or 10% zirconium oxide are needed to opacify the glaze. Titanium dioxide can be used as an opacifier but may produce a cream colour on stoneware.

Tin oxide (SnO_2) was the original opacifier used by potters but it has now become very expensive. Tin was once mined in Cornwall but now comes from Southeast Asia and South America. Around 5–10% tin oxide is needed to opacify a glaze. The melting point of tin oxide is 1150°C (2102°F) and a small amount (1–2%) dissolves in the glaze and can act as a flux at stoneware temperatures. Tin oxide is an essential ingredient in both chrome-tin pink glazes and copper reds, where it helps to stabilise the colour. However, white tin glazes may pick up pink flashing from other pots in the kiln. Tin oxide produces a soft, milky white.

A less expensive opacifier is zirconium silicate ($ZrSiO_4$), although more (10–15%) is needed to fully opacify a glaze. Zirconium silicate has a very high melting point of 2550°C (4622°F). It gives a cold, hard white which may sometimes be marked by cutlery. However, zirconium will not pick up pink flashing from copper (in reduction) or chromium (in oxidation) during firing in the kiln. Zirconium silicate gives a glassy, opaque finish.

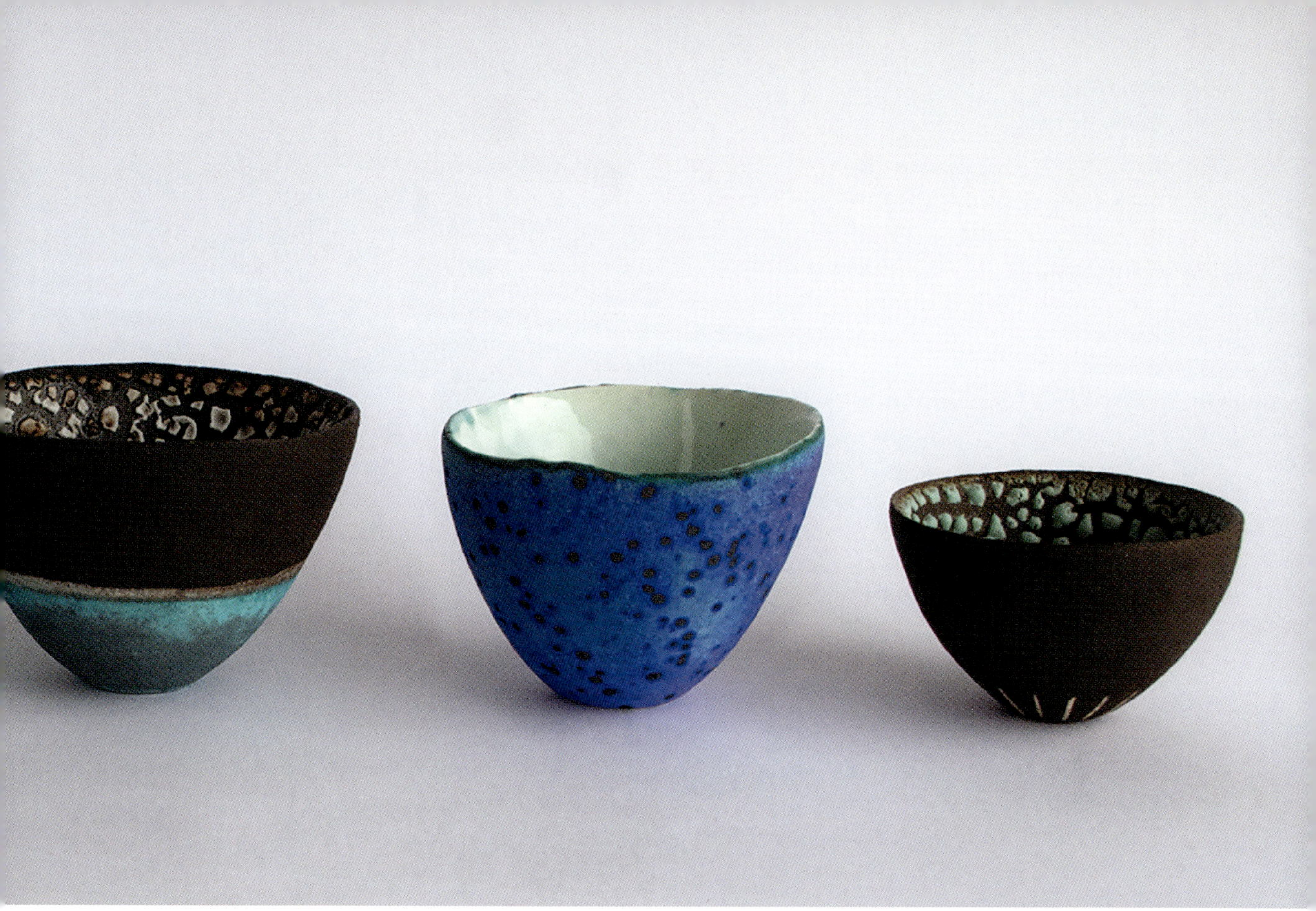

ABOVE: Emma Williams, Tall bowls, black clay, barium glazes, fired to 1055°C (1931°F).

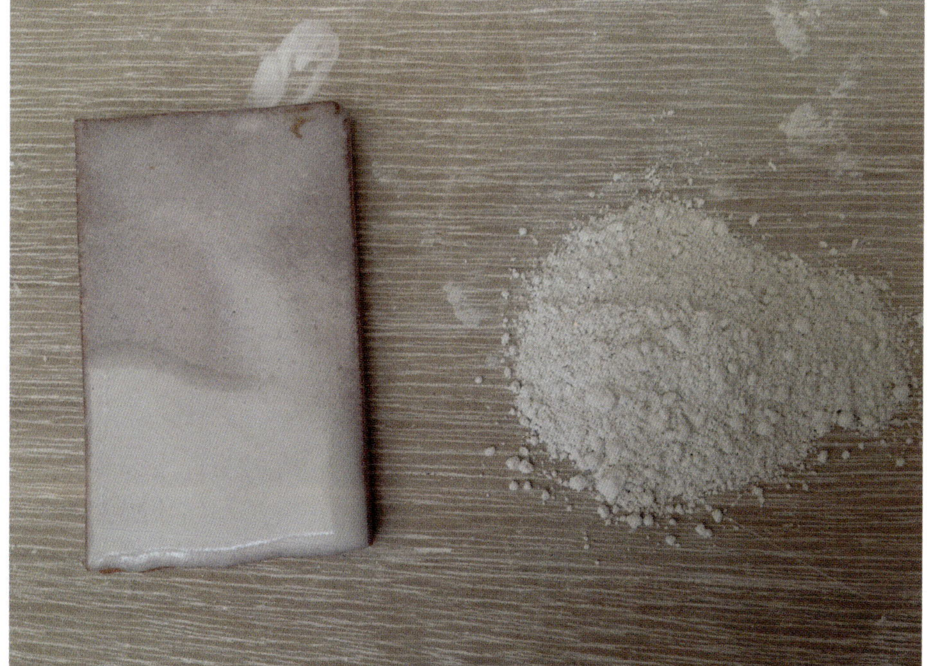

Zirconium-opacified glaze on red earthenware . Glaze: calcium borate frit 39, soda feldspar 27, whiting 5, china clay 6, quartz 23, zirconium silicate 5 (fired to cone 04). The powder is tin oxide.

Titanium dioxide (TiO_2) is used to encourage mottling and crystallisation, but it also makes glazes opaque. 5–10% can be used in a glaze to give opacity and mattness. However, it tends to combine with iron oxide from the clay body. Titanium is also found together with iron oxide in ilmenite (50% iron oxide) and rutile (up to 15% iron oxide). The presence of iron oxide gives it a beige or tan colour, so on buff stoneware it is a less white opacifier than zirconium or tin. The melting point of titanium dioxide is 1830°C (3326°F). As it remains unmelted, it can act as a seed material for crystals to grow in low alumina glazes.

Opacifiers are refractory materials which can be fired to a high temperature without melting. They remain as white particles suspended in the molten glaze, causing it to become opaque. Because zirconium is an acidic oxide, it acts as an anti-flux, and can cause the glaze to become underfired. If this happens, you can reduce the amount of silica (quartz or flint) in the glaze by 5% or add a small amount (1–2%) of zinc oxide.

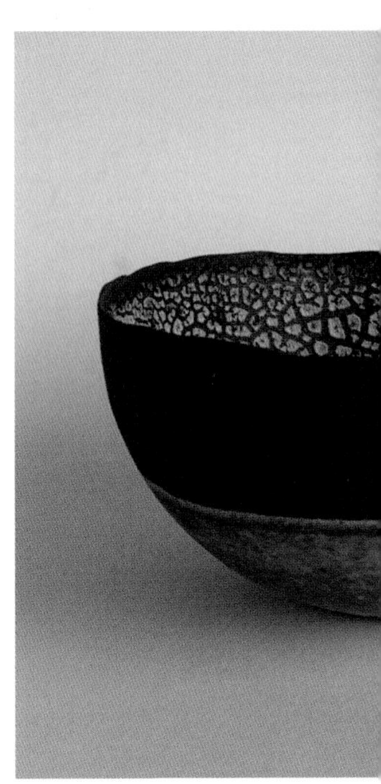

Silicon carbide

Silicon carbide (0.2–2%) is added to glazes either as coarse granules (60–180 mesh) or as a fine powder (360–1200 mesh) to cause local reduction of copper reds and celadons and to cause craters in volcanic glazes. If a large quantity (over 5%) is used, it can cause frothing and may colour the glaze grey.

Stains and underglaze colours

Commercial stains are made by heating colouring oxides together with silica and opacifiers and grinding to a powder. Many stains contain zirconium silicate, which makes the glaze opaque. If you want to make a bright yellow or red glaze, you will need to buy stains, as those colours cannot be made easily using oxides. Red and orange inclusion stains, where the pigment is encapsulated in a matrix of zirconium silicate, are stable up to high temperatures. If you want blue, green or brown it is better to use oxides, which give more interesting depth of colour than stains. Some potters prefer to use black stain rather than oxides, as the former gives a truer black. Underglaze colours are stains mixed with clay and frit to make them easier to apply.

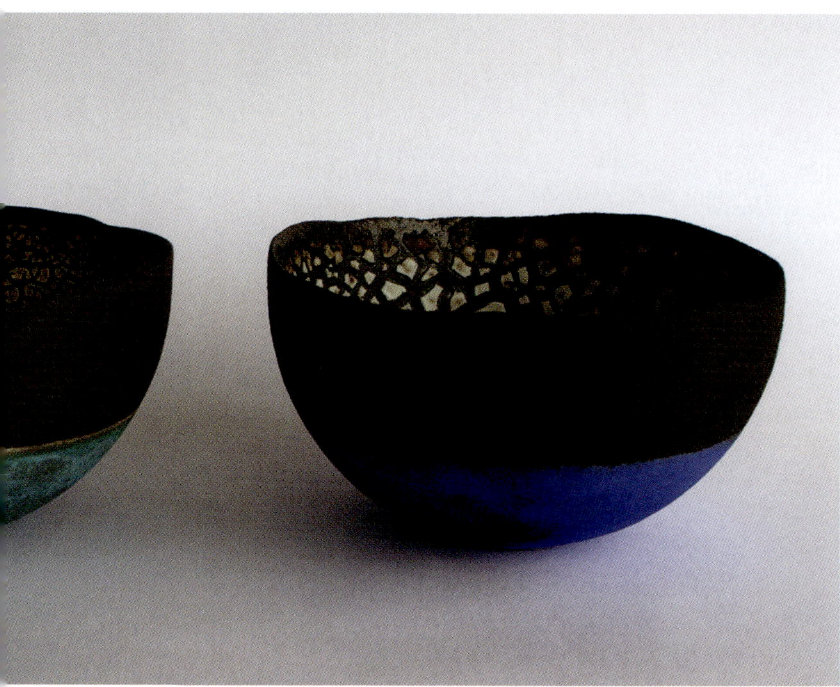

ABOVE: Emma Williams, black earthenware bowls with white crackle glaze on the inside and barium matt glaze on the outside. The blue is the same glaze used by Emily Myers.

Coloured glaze recipes

Matt Blue (Emily Myers), cone 04 (1050°C/1922°F).

(Not food-safe)

Barium carbonate 40
China clay 19
Nepheline syenite 19
Flint 10
Lithium carbonate 5

+ copper carbonate 3.5 vibrant purple
+ cobalt carbonate 3 deep blue

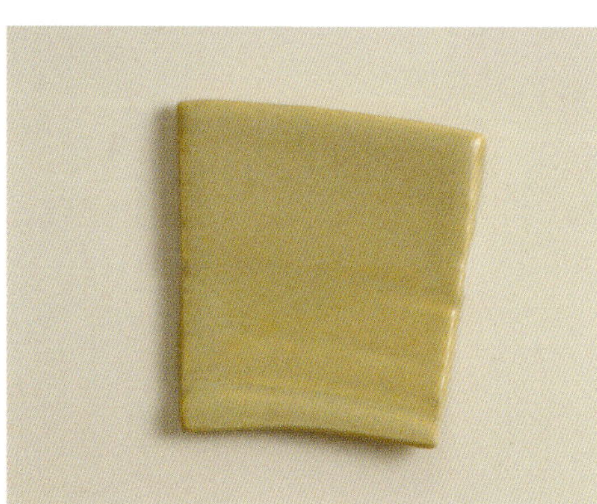

Satin matt mustard glaze, cone 8 (1250°C /2282°F).

Potash feldspar 33
Talc 21
Whiting 12
China clay 15
Quartz 16
Zinc oxide 3
+
Nickel oxide 3
Titanium dioxide 10

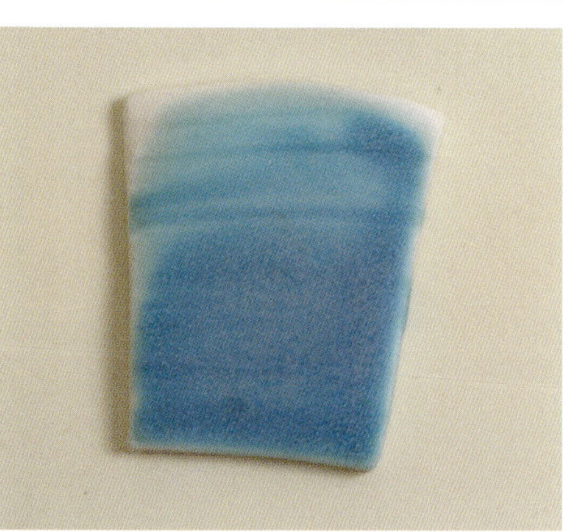

Strontium matt turquoise glaze, cone 8 (1250°C/2282°F).

(Not food-safe)

Nepheline syenite 60
Strontium carbonate 21
Lithium carbonate 2
China clay 6
Flint 9
Calcium borate frit 2
+
Copper oxide 2

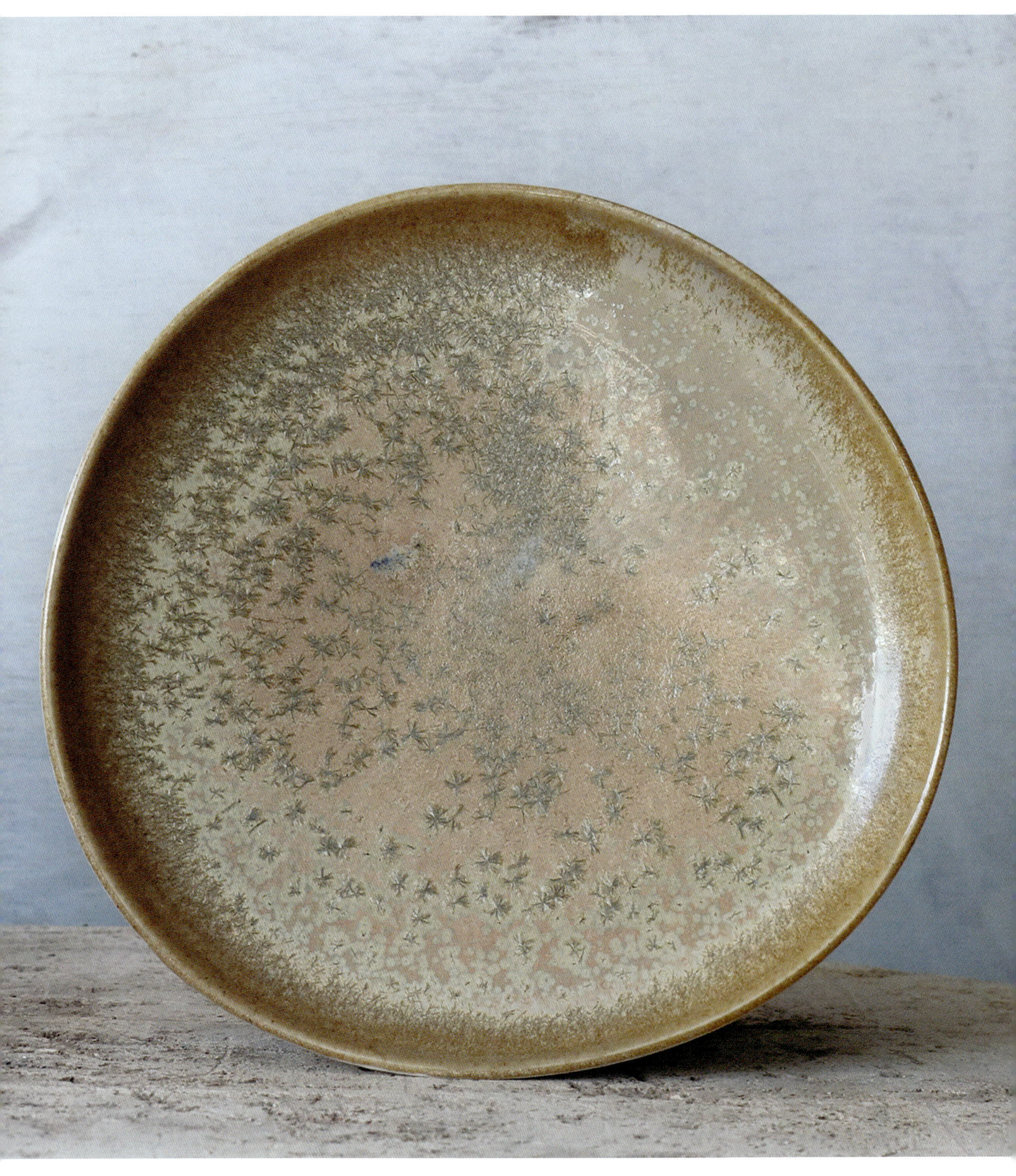

4

Impurities and variation in materials

Many potters use locally found materials such as clay, naturally occurring slips, wood ash and quarry rocks. The interesting effects they produce often come from impurities in the materials such as iron oxide, titanium dioxide and other trace elements. Many glaze materials purchased from potters' suppliers also contain impurities: feldspar, ball clay, and rutile are among the most variable. It is interesting to discover more about the kind of impurities present and how they affect the glaze. Here, as an example of this natural variation, we will investigate the impurities found in rutile.

Rutile is a naturally occurring mineral that is used as a source of titanium dioxide to colour and opacify glazes. It is mined from heavy mineral deposits in beach sands in Australia, South Africa and Sierra Leone. Ilmenite, $FeTiO_3$ contains more iron oxide (50%), while rutile contains less than 15% iron oxide and can be purified further to white titanium dioxide TiO_2. Rutile can give a variety of effects in glazes, including streaking, mottling and opalescence. In reduction firing, it produces a blue effect. This is thought to be caused by scattering of light by small particles of rutile suspended in the glaze. It could also be caused by electron transfer between reduced Ti^{3+} and Ti^{4+} ions, which absorbs yellow light, causing the remaining light to appear blue. A similar charge transfer process causes sapphire to appear blue.

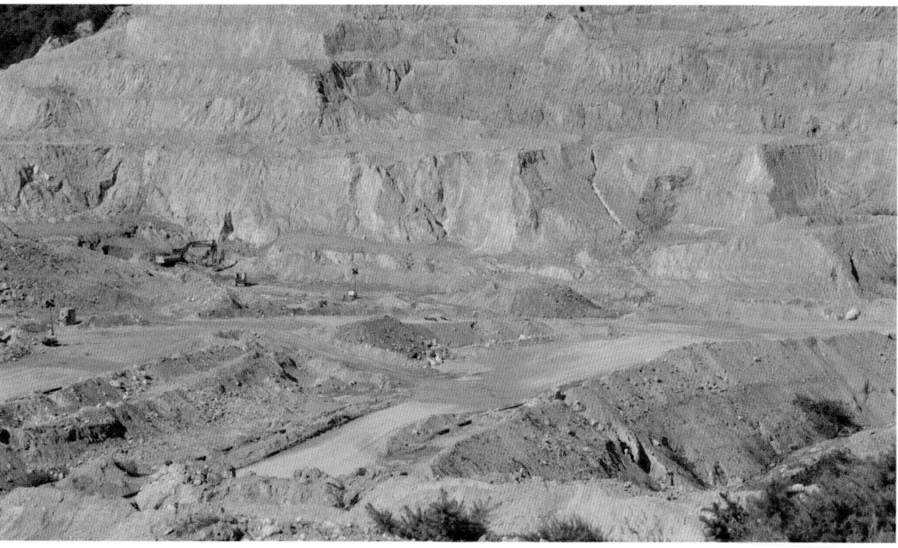

ABOVE: Clay pit at St Austell, Cornwall, UK.

LEFT: Golden plate, porcelain, crystal glaze coloured with UK rutile 12% and red iron oxide 8%, dia: 21.5cm (8¼ in.), from the *Wilder* range by Pottery West. *Photo: Jules Lister for Labrador 2017.*

Titanium dioxide on its own produces opacity but not colour in glazes fired in oxidation (see image d, left tile, opposite). However, rutile also contains up to 15% iron oxide as part of its crystal structure and additional impurities including chromium, vanadium and niobium. The concentration of impurities is low (light rutile can contain less than 1% iron oxide) but can still have an effect on the colour of the glaze. The iron oxide causes rutile to colour glazes an ochre yellow in oxidation. The base glaze must be relatively high in alumina (clay) for this tan yellow colour to appear (image d). The small amount of chromium in rutile can cause a pale pink colour to appear in glazes containing tin oxide (image c). Instead of dissolving in the glaze to give the usual green colour (from Cr^{3+} ions), the chromium reacts with the tin oxide, calcia and silica to form a crystal structure known as sphene, or calcium tin silicate ($CaO.SnO_2.SiO_2$ which has Cr^{4+} ions substituted for some of the tin). In order for this reaction to occur, there must be sufficient calcium and silica in the glaze. The pink colour can be obtained from rutile if the glaze is low in alumina, magnesia (dolomite or talc), and also contains no zinc oxide. The pink is paler than that obtained using pure chromium oxide, which tends to give tiny specks of a darker, maroon colour. This is because the rutile contains less than 0.1% chromium, so the particles of chromium oxide are more finely divided in rutile than in pure chromium oxide. Thus we can see that these minor components of rutile can have subtle effects on the resulting glaze colour.

I set up my first pottery studio while living in California, before moving back to to London. After testing various rutile-tin pink glaze recipes (image c), I noticed that the rutile obtained from potters' suppliers in the US gives a purer pink than rutile obtained in the UK. I had the two rutile samples analysed and found that the UK sample contained more iron oxide. This could account for the orange-tinged peach colour obtained using UK rutile. I had thought vanadium might also be contributing to the colour, but the UK sample had lower vanadium than the US sample so this is apparently not the case. The vanadium in rutile does not have much effect on glaze colour as vanadium is not a strong glaze colourant – i.e. a larger amount (5–10%) of vanadium pentoxide would be needed to colour a glaze yellow.

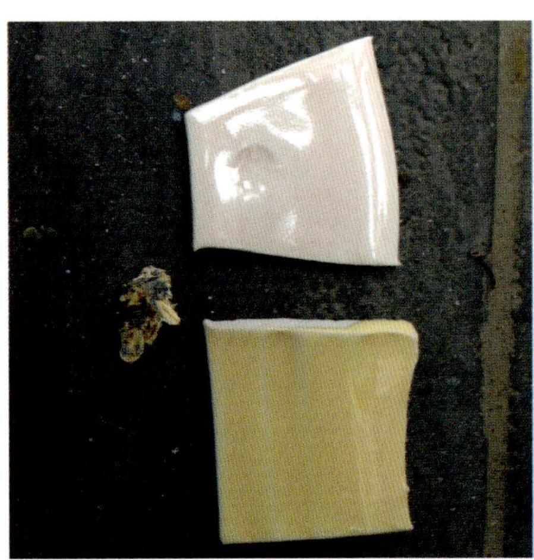

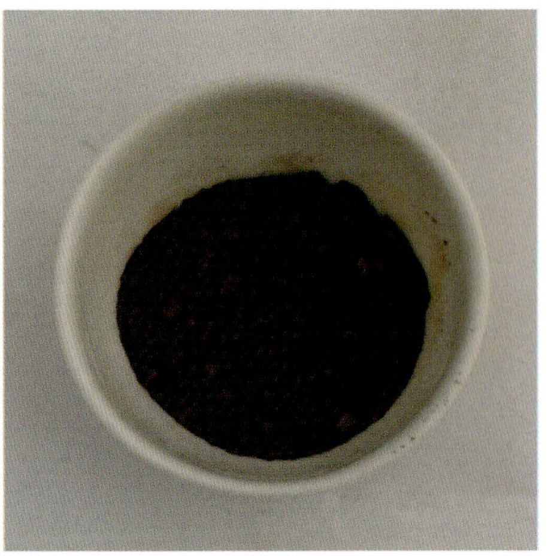

Powdered rutile: a. impure TiO_2 (light brown), and b. ilmenite $FeTiO_3$ (dark brown).

Glaze recipes

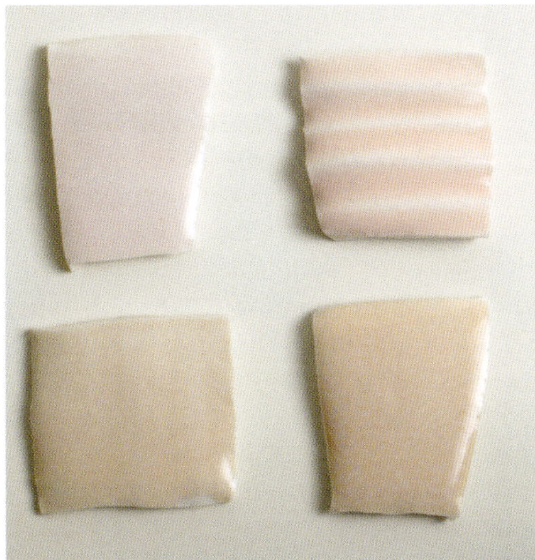

c. Rutile-tin pink glaze tests on porcelain using UK and US rutile, fired to 1250°C (2282°F) in oxidation; top, US rutile-tin pink, bottom UK rutile-tin pink.

Rutile-tin pink, cone 8, 1250°C (2282°F), oxidation, (left tile thin glaze, right tile thick glaze).

Soda feldspar 47
Quartz 18
Calcium borate frit 15
Whiting 14
China clay 5
+
Tin oxide 4
Rutile 2

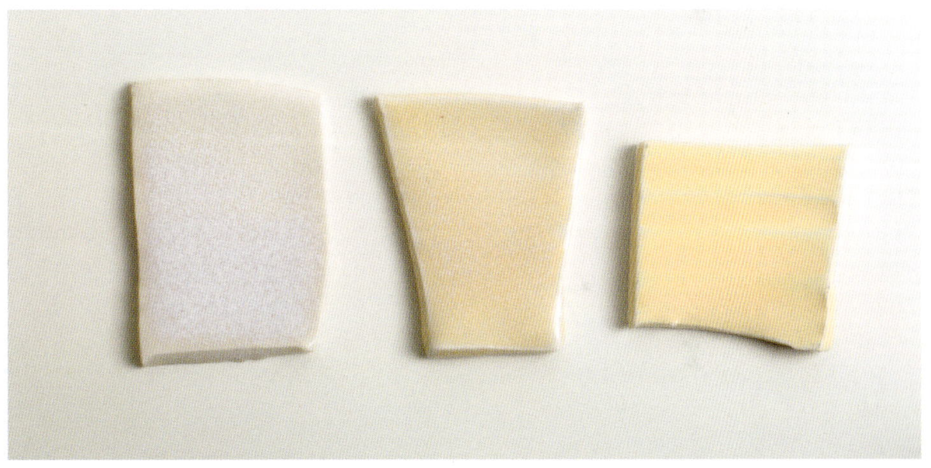

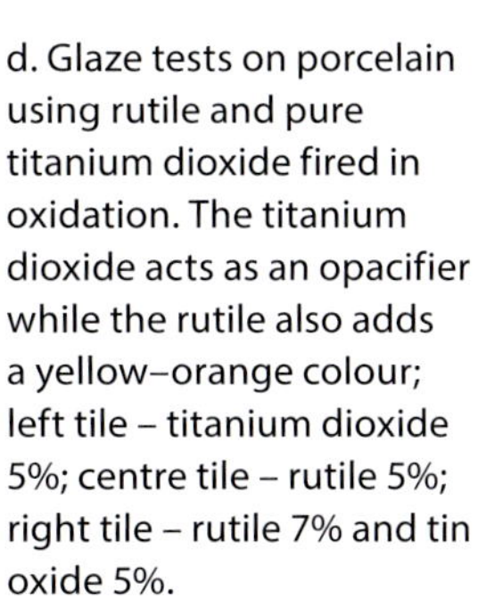

d. Glaze tests on porcelain using rutile and pure titanium dioxide fired in oxidation. The titanium dioxide acts as an opacifier while the rutile also adds a yellow–orange colour; left tile – titanium dioxide 5%; centre tile – rutile 5%; right tile – rutile 7% and tin oxide 5%.

Rutile tan yellow, cone 8, 1250°C (2282°F), (image d, centre tile).

Potash feldspar 34
Quartz 23
Calcium borate frit 14
China clay 13
Whiting 11
Dolomite 5
+
Rutile 5

Rutile yellow, cone 8, 1250°C (2282°F), (image d, right tile).

Potash feldspar 33
Talc 21
Whiting 12
China clay 15
Quartz 16
+
Tin oxide 5
Rutile 7

When we buy materials from potters' suppliers, we can also obtain a Materials Safety Data Sheet (MSDS) with details of any toxicity (rutile is not toxic apart from dust from the silica content). The Materials Safety Data Sheets for rutile state that it contains more than 90% titanium dioxide, and less than 0.5% chromium oxide, as well as zircon and silica. I had been thinking of having my rutile samples analysed for some time in order to be able to reproduce my glaze colours in case the current supplies of rutile run out. I recently discovered that analysis is readily available and is not expensive. I would recommend this method of analysis to potters using locally found materials if they are interested in the composition of the material and the possibility to reproduce the glaze should the material run out.

The rutile samples were anaylsed by Wheal Jane Laboratory, which is based on the site of an old tin mine in Cornwall. Two techniques were used to detect trace elements: x-ray fluorescence (XRF) and interactively coupled plasma (ICP) optical emission spectrometry. This sounds complicated but the latter is basically a more precise version of the test we do in chemistry lessons at school when we heat metal compounds to see what colour the flame is (e.g. sodium compounds have a yellow flame, lithium red, barium pale green, as used to make coloured fireworks). In x-ray fluorescence spectroscopy, the sample is bombarded with x-rays and the colours of light emitted are analysed. The x-rays knock out electrons from the sample atoms and other electrons fall into the holes left, releasing energy in the form of light. Each element present has signature colours of light emitted.

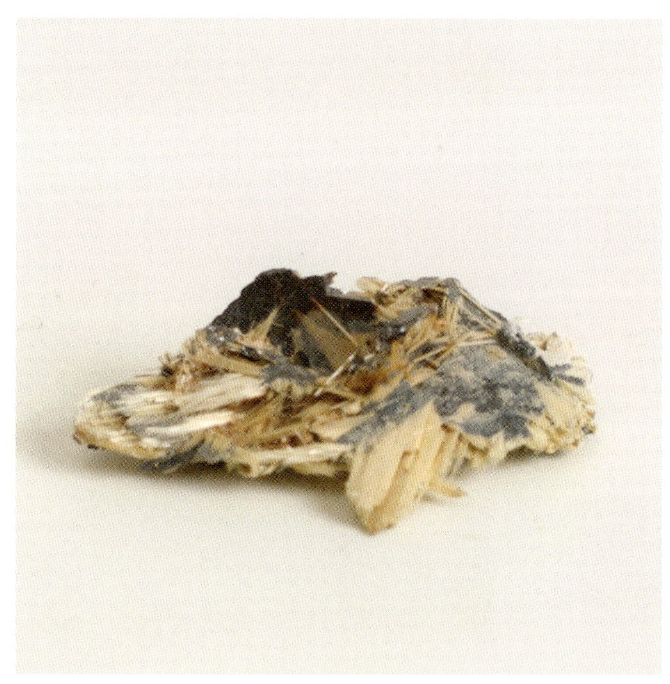

ABOVE: Rutile mineral, with needle-shaped crystals.

FAR RIGHT: Golden plate, porcelain, crystal glaze coloured with UK rutile 12% and red iron oxide 8%, dia: 21.5cm (8½in.), *Wilder* by Pottery West. *Photo: Jules Lister for Labrador 2017.*

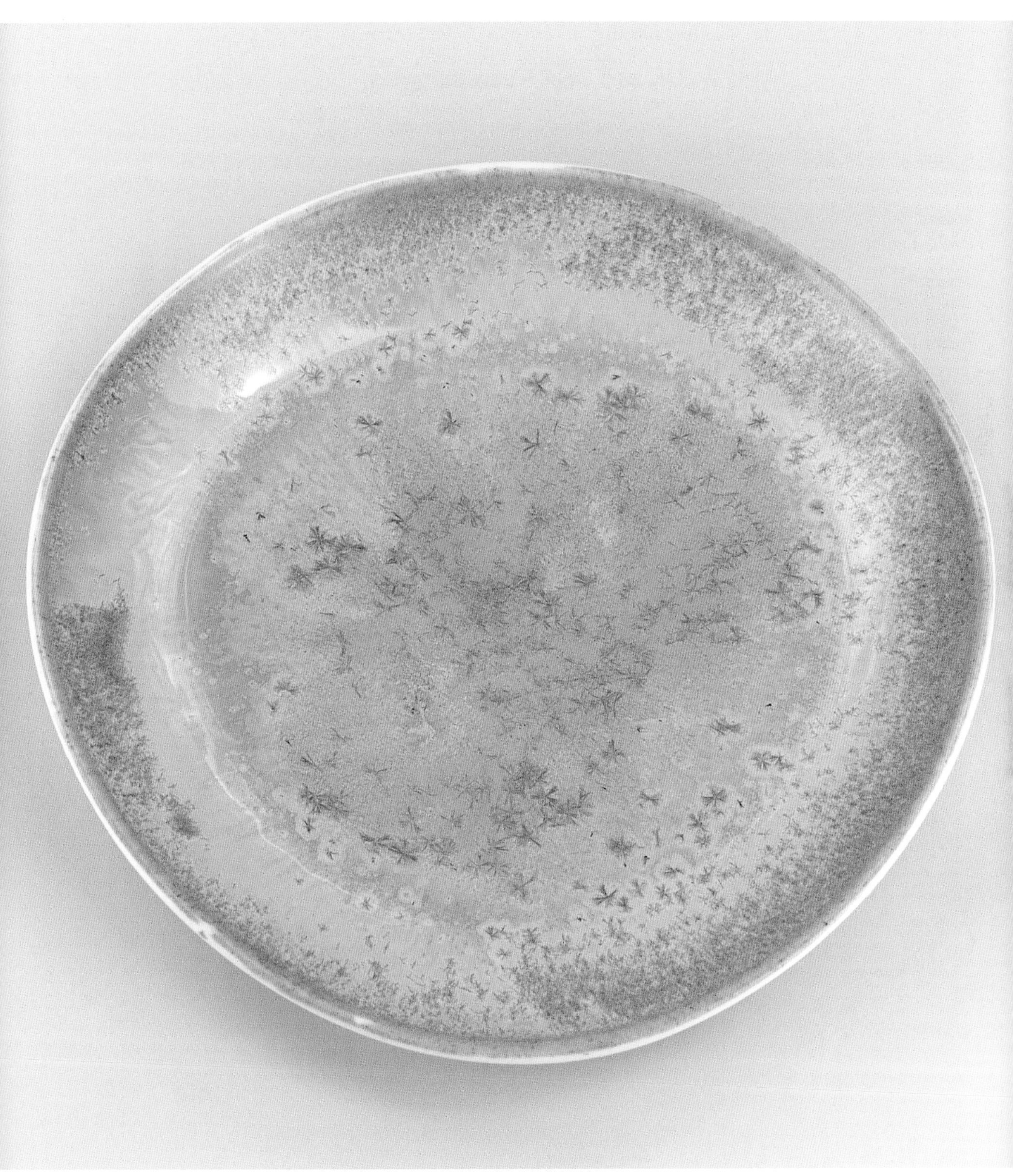

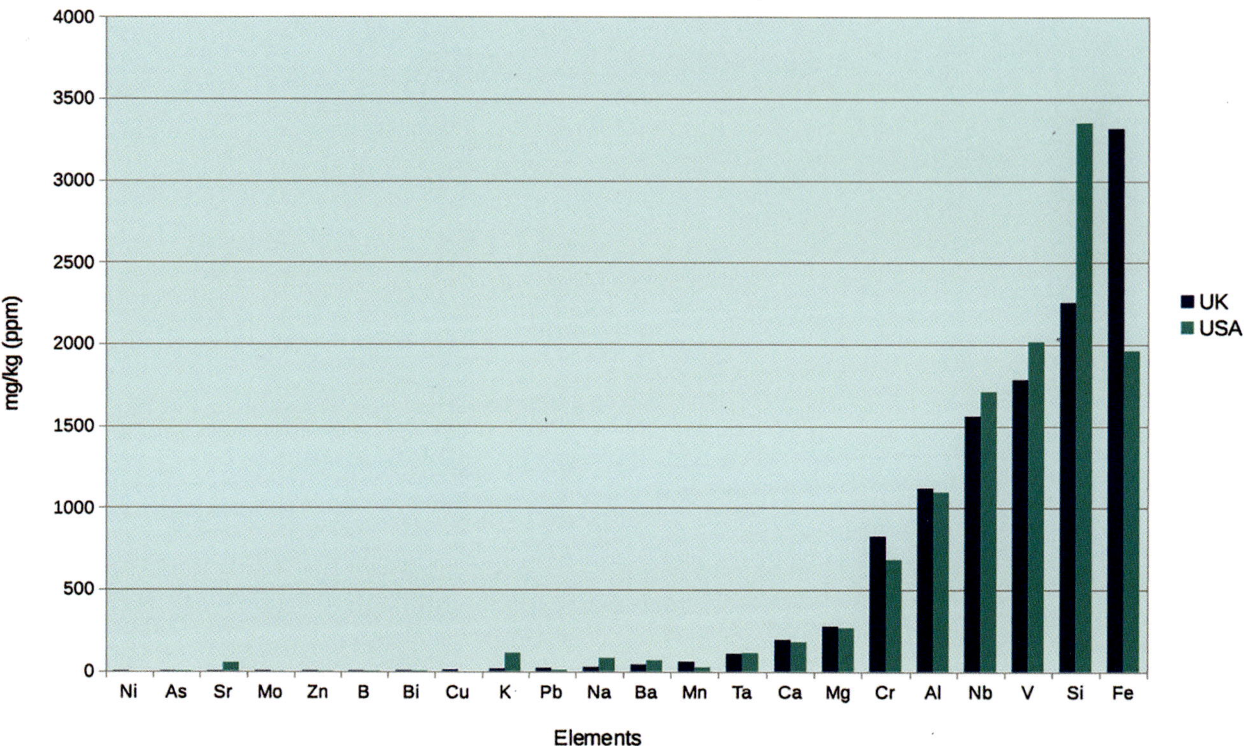

In the bar chart, the concentration of impurity elements is shown in milligrams per kilogram or parts per millon (1000 ppm = 0.1%). The samples also contained more than 1% zirconium and 90% titanium, not shown on the chart. The elements Si, Al, Mg, Ca, Ba, Na, and K would have little effect on the colour and are already present in many glazes. The elements that have an effect on colour are the transition metal elements iron Fe, vanadium V, chromium Cr, and manganese Mn. However, vanadium and manganese are relatively weak colourants, which leaves iron and chromium as the most important colouring impurities present. The results show that the UK rutile contained more iron and chromium than the US rutile. In glazes, iron oxide imparts a range of ochre, tan or brown colours in oxidation, while chromium oxide is green, or pink when combined with tin oxide in a glaze. The titanium and zirconium in the rutile contribute to opacity when more than 2% rutile is added to a glaze. These oxides have high melting points and stay as unmelted particles in the glaze, causing opacity. The elements niobium Nb and tantalum Ta are unusual metals whose oxides also contribute opacity and opalescence to the glaze.

Graph showing results of trace element analysis of UK and US rutile samples. The major impurities include iron, vanadium, niobium and chromium, as well as zirconium and silicon.

Theoretically composed of titanium oxide and iron oxide, rutile supplied to potters actually contains a variety of different impurities including iron, chromium, vanadium and niobium. UK rutile contains more iron and chromium, while US rutile contains slightly more vanadium and niobium. The UK rutile comes from Australia and was supplied by CTM potters' supplies. The US rutile came from Ferro Electronic Materials, Niagara Falls, New York. According to Tony Hansen of Digitalfire.com, this US rutile is blended from rutile sourced from Australia and Florida. The difference detected may be due to the Florida rutile, or different Australian rutile supplies.

Other colouring oxides such as cobalt, copper and chromium oxide are relatively pure, as are single mineral oxides such as quartz, and silicates talc and wollastonite. However, many clays and feldspars come as dug from the ground and are less purified, often containing impurities such as iron oxide and titanium dioxide, which can affect the colours of glazes. Materials sourced locally by potters such as wild clays or wood ash may have many impurities that can enhance the appearance of the fired clay and glaze (see chapter on wood ash, p.89).

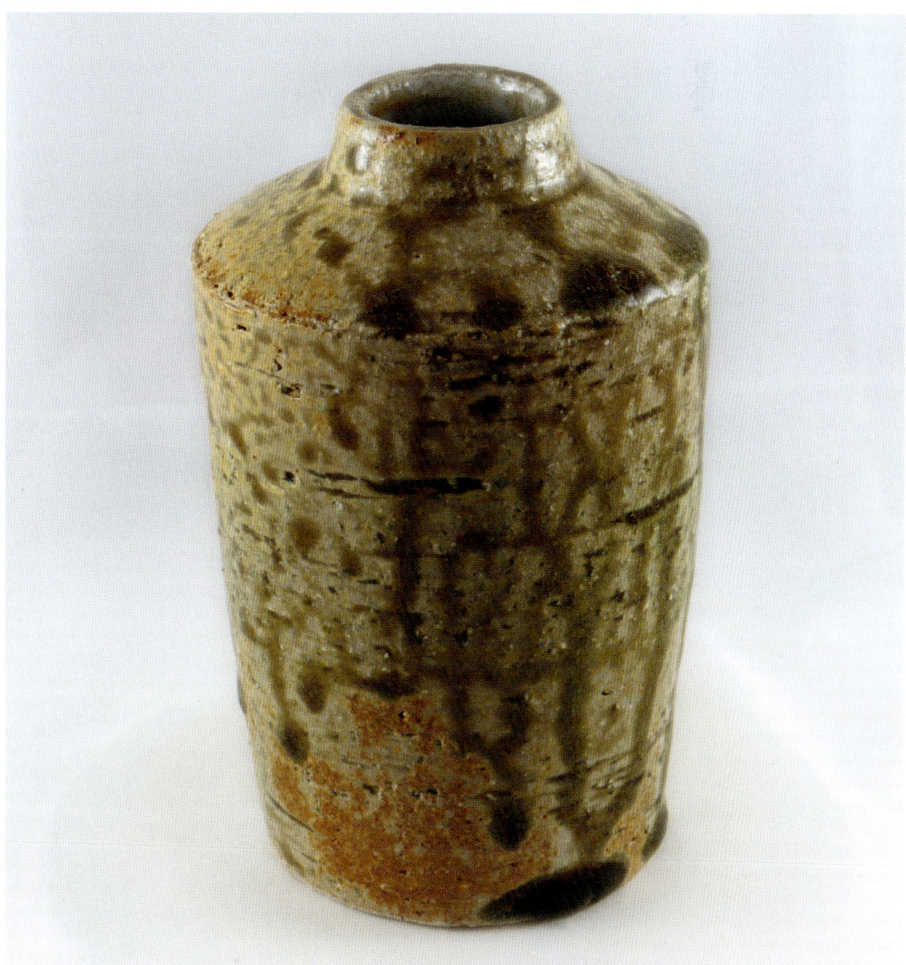

Robert Hunter, olive ash-glazed stoneware bottle, fired in reduction. In ash glazes the green colour comes from iron oxide in the wood ash.

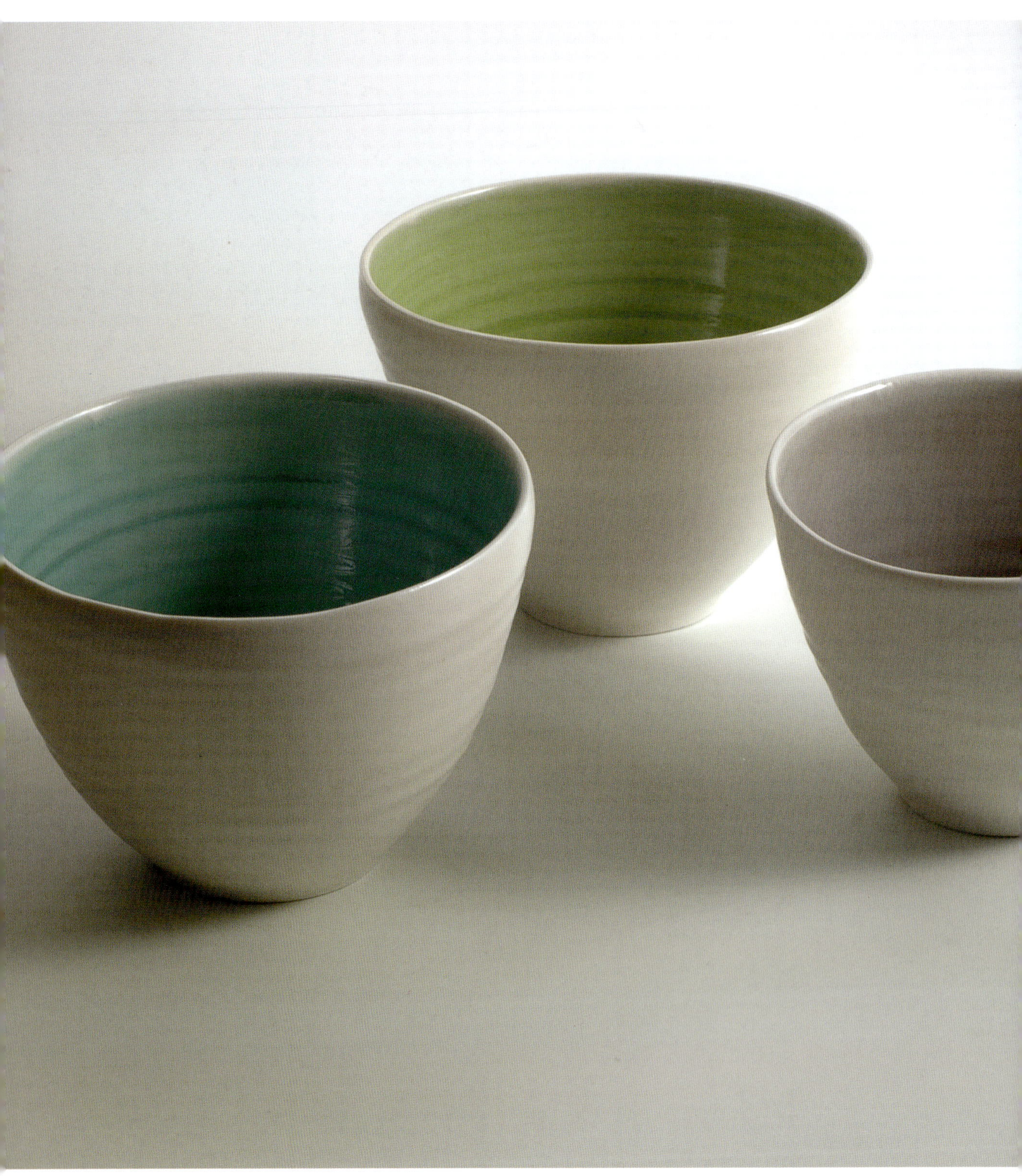

5

Stability and durability

Most of the time, when we think about creating or modifying a glaze, we are thinking of its colour, texture, or how runny it is, but other aspects of glazes such as how closely the thermal expansion of the glaze matches that of the clay body or how stable and durable it is are just as important. When we look in detail at special-effect glazes later in this book, we will see that many of these glazes push the boundaries of what is 'normal' in glazes, and an understanding of how to control them helps you to get the most from these effects.

In order to understand the stability and durability of a glaze, we need to go back to the basics of glaze formulation. Glazes are made from three components: silica – the glass former; fluxes to melt the silica; and clay to stiffen the glaze when melted. These components need to be combined in the right proportions to achieve the best melting at the chosen kiln temperature. The choice of fluxes used is also important and affects the glaze durability. It is better to use a combination of several fluxes to achieve melting. The most powerful fluxes are the alkali metal oxides of potassium (K), sodium (Na) and lithium (Li). Their oxides have the chemical formula R_2O where R is the alkali metal K, Na or Li and O is oxygen. They help the silica to melt at a lower temperature than its natural melting point of 1710°C (3110°F). However, sodium silicate is soluble in water so would not make a good glaze on its own. To make the glaze insoluble, an alkaline earth such as calcium must be added. The alkaline earth oxides of calcium (Ca), magnesium (Mg), barium (Ba) and strontium (Sr) are less powerful fluxes than the alkali metals but help to stabilise the glaze, making it more resistant to attack by water, acid or alkali. Their oxides have the chemical formula RO where R is the alkaline earth Ca, Mg, Ba or Sr, and O is oxygen. Low-temperature fluxes lead (Pb) and zinc (Zn) also have the formula RO.

There are exceptions to this rule, including shino glazes, which only contain alkali metal fluxes (made from feldspar and clay), and soda- or salt-glazed ware, which is glazed by introducing sodium salts into the kiln at high temperatures. The sodium becomes volatile and reacts with the clay surface, forming a sodium alumino-silicate glaze. These types of glazes are fired at high temperatures (1280–1300°C/2336–2372°F) and the high levels of alumina and silica in the glaze ensure all the alkali metals are safely tied up in the glaze structure.

Linda Bloomfield, thrown porcelain bowls with runny turquoise glaze, chromium green and nickel–cobalt grey with satin matt glaze on the outside.

Stable glazes

The proportion of alkali metal to alkaline earth is very important when considering the chemical stability of the glaze. The ratio of 0.3:0.7 R_2O:RO has been found to make the most stable glaze; resistant to attack by acid in food and alkali in dishwasher soap. If there is excess alkali metal oxide in the glaze, it will not not be tied up in the glaze structure and it will be likely to leach in acids and corrode in the dishwasher. The excess alkali metal atoms can exchange with the hydrogen ions (H^+) released by acids. Hydrogen ions are much smaller than alkali metal ions, so will leave holes in the glaze structure that greatly weaken the glaze and cause it to break down (see diagrams). The same process causes fluorine in some clays to etch the windows of the pottery studio when it becomes volatile during firing and forms hydrofluoric acid in contact with water. Acids found in foods such as vinegar and lemon juice can cause toxic metals in an unbalanced glaze to leach into the food.

The alkali in dishwasher soap supplies hydroxyl ions (OH^-) which attack the silica in the glaze and cause the glaze surface to dissolve if it is not chemically stable. The same process causes glassware to etch in the dishwasher. Dishwasher soap can be very caustic, having a pH between 11 and 13 (the pH scale goes up to 14). The scouring action of the hot water and alkaline soap can be very destructive to glazes.

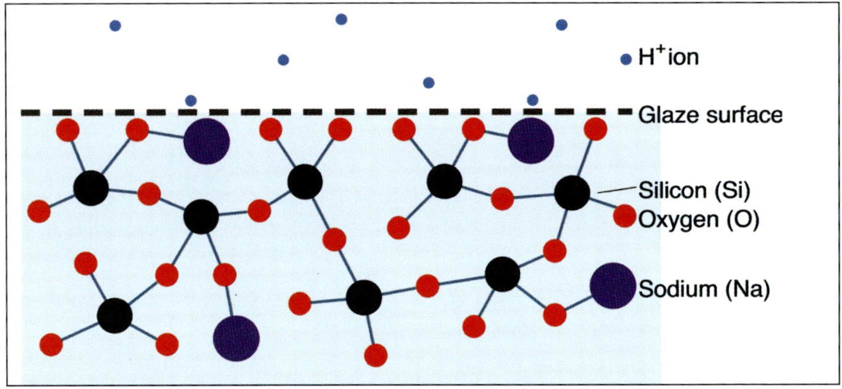

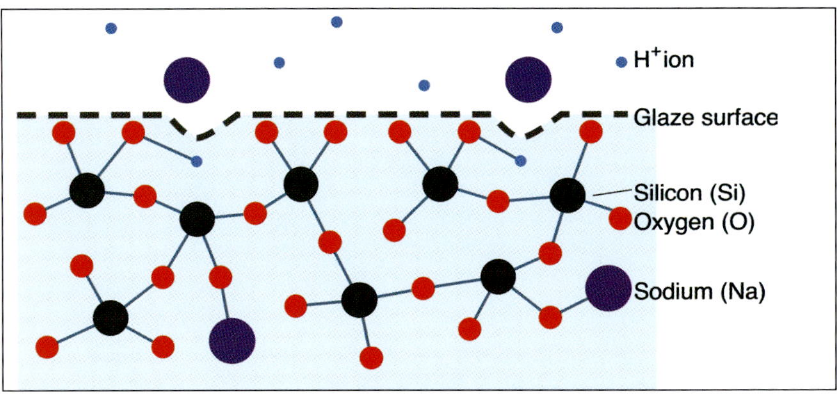

These diagrams show what happens to an unbalanced glaze in contact with an acid. Acids release hydrogen H^+ ions, which can exchange with excess sodium or potassium ions in the glaze. As the hydrogen ions are smaller than the sodium ions, this ion exchange process leaves holes in the glaze surface. The opposite process, using larger potassium ions, is used to compress and toughen the surface of soda-lime glass in mobile phone screens. Diagrams by Henry Bloomfield.

Glaze calculation

To determine how much alkali metal oxide is present in a particular glaze, the glaze recipe is entered into a glaze calculation program such as the one on glazy.org, where the molecular formula of a glaze can be calculated (see below). This will determine the molecular ratio of alkali metal oxide to alkaline earth oxide R_2O:RO. A ratio of 0.3:0.7 is ideal, but the glaze will also be chemically stable and durable at 0.2:0.8 or 0.4:0.6 R_2O:RO. Glazes become unbalanced and much less durable if the alkali metal is increased to 0.5:0.5 or even 0.7:0.3 (see the high-alkaline turquoise crackle glaze recipe on p.47). This glaze will eventually take on a sandblasted, matt appearance if washed in the dishwasher over a large number of cycles (for example, every day for two months). However, some examples of pottery from ancient Egypt and Persia glazed with this type of turquoise alkaline glaze have lasted several thousand years.

Calculating the unity molecular formula

Glaze recipes can be written either as a recipe by weight percent, or as a chemical formula. To calculate the formula, each material in the glaze recipe is converted into a number of molecules, according to their relative proportions in the glaze recipe. To make this conversion, the weight of each material is divided by its molecular weight (see Appendix 3). Dividing by the total number of molecules of alkaline fluxes (sodium, potassium, calcium, etc.) gives the unity molecular formula (UMF), where the sum of the alkaline fluxes is arbitrarily set to 1. This provides the ratios of alumina to silica and alkali metal to alkaline earth oxides, which determine whether the glaze is glossy, matt or likely to be durable. Online glaze calculators such as glazy.org can be used to calculate the molecular formula for a glaze recipe.

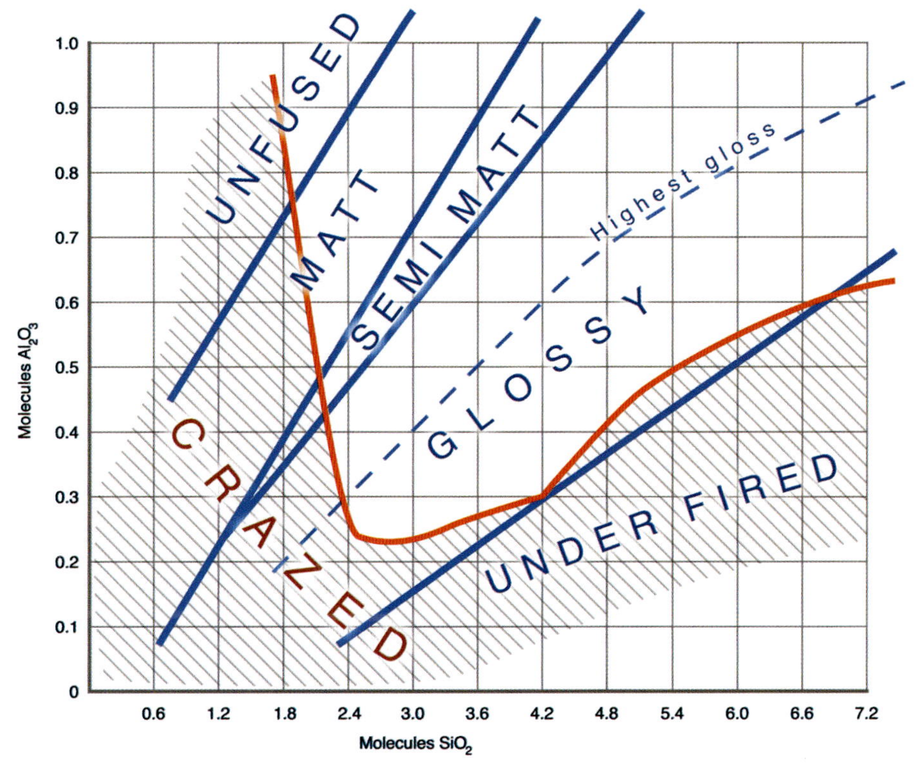

Graph of alumina against silica in porcelain glazes fired to cone 11 with constant flux 0.3 K_2O and 0.7 CaO. The ratio of 1:5 alumina to silica gives a semi-matt glaze, while 1:8 gives a shiny glaze. The straight lines on the chart represent alumina:silica ratios of 1:4 (matt), 1:5 (semi-matt) and 1:12 (glossy, crazed glaze). The dashed line is 1:8 Al_2O_3:SiO_2 (highest gloss glaze). The hatched area shows crazed glazes on porcelain. Data from R.T. Stull 1912.

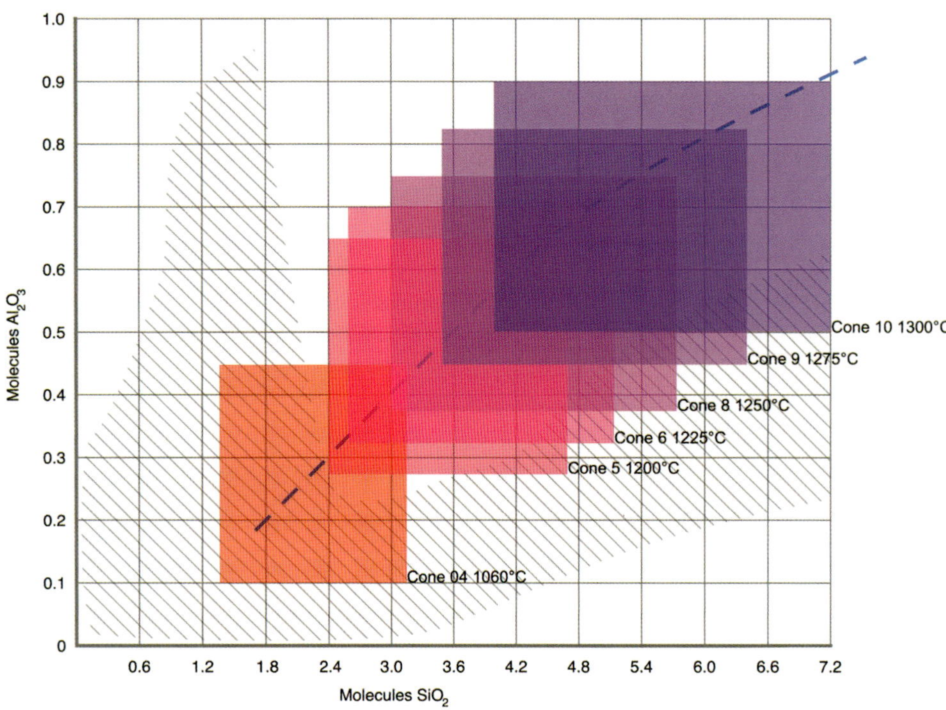

Graph of alumina against silica showing the limits for stable glazes set out in the table below. The higher the firing temperature, the more silica and alumina can be added and the stronger and harder the fired glaze will be. These limits are mostly inside the area for glossy glazes. (See p.150 for cone conversion temperatures.)

Limits for stable glazes:

Alumina and silica (Cooper and Royle, 1984)

Cone number and temperature.		Number of molecules in unity formula.			
Cone 04	1060°C (1940°F)	Al_2O_3	0.1–0.45	SiO_2	1.375–3.15
Cone 5	1200°C (2192°F)	Al_2O_3	0.275–0.65	SiO_2	2.4–4.7
Cone 6	1225°C (2237°F)	Al_2O_3	0.325–0.70	SiO_2	2.6–5.15
Cone 8	1250°C (2282°F)	Al_2O_3	0.375–0.75	SiO_2	3.0–5.75
Cone 9	1275°C (2327°F)	Al_2O_3	0.45–0.825	SiO_2	3.5–6.4
Cone 10	1300°C (2372°F)	Al_2O_3	0.50–0.90	SiO_2	4.0–7.2

Recommended maximum flux in glaze unity molecular formula.
(Cooper and Royle, 1984)

Cone	Temp °C	MgO	BaO	ZnO	CaO	B_2O_3	K+Na
5	1200°C (2192°F)	0.325	0.40	0.30	0.55	0.35	0.375
6	1225°C (2237°F)	0.330	0.425	0.32	0.60	0.30	0.35
8	1250°C (2282°F)	0.335	0.45	0.34	0.65	0.25	0.325
9	1275°C (2327°F)	0.340	0.475	0.36	0.70	0.225	0.30
10	1300°C (2372°F)	0.345	0.50	0.38	0.75	0.21	0.275

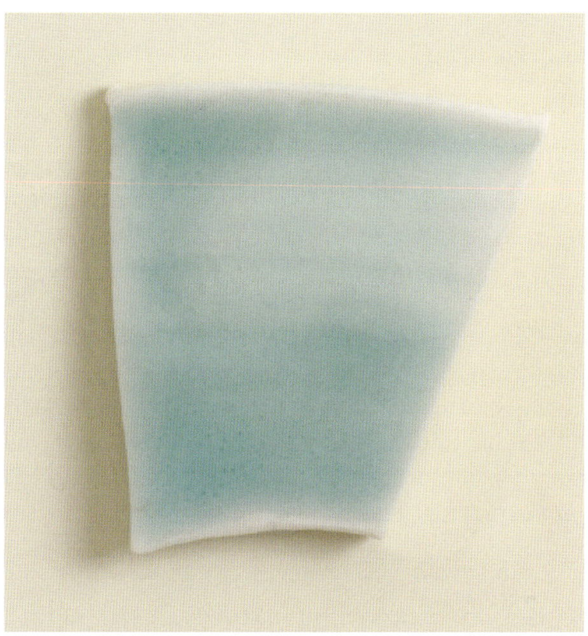

Runny turquoise glaze, cone 6–8, 1240–1260°C (2264–2300°F).
This glaze has a low ratio of sodium to calcium and will be relatively durable.
(Unity molecular formula: Na_2O 0.25, CaO 0.75, B_2O_3 0.38, Al_2O_3 0.38, SiO_2 3.1)

Soda feldspar 47
Calcium borate frit 16
Whiting 14
China clay 5
Quartz 18
+
Copper oxide 1

High-alkaline turquoise crackle glaze, cone 6, 1240°C (2264°F).
This glaze has very high sodium and is unlikely to be very durable.
(Unity molecular formula: Na_2O 0.7, CaO 0.3, Al_2O_3 0.23, SiO_2 2.5). Not food-safe.

Soda feldspar 15
High alkaline frit 47
Lithium carbonate 2
Whiting 6
Quartz 18
China clay 10
+
Copper oxide 2

Glaze materials

Alkali metal oxides are supplied in glazes mainly by soda and potash feldspars, or nepheline syenite, a more concentrated source of sodium than feldspar. You could also use lithium carbonate, soda ash (sodium carbonate) or pearl ash (potassium carbonate). However, soda ash and pearl ash are soluble in water, so are not often used in glazes as they can cause problems. These materials can also be supplied by alkaline or borate frits, which are insoluble in water. Beware of any glaze recipes with more than 10% lithium carbonate, as these may not form chemically stable glazes, owing to the high proportion of alkali metal (lithium carbonate is a much more concentrated form of lithium than a lithium feldspar). It is better to use a frit containing boron if you want to lower the melting temperature of your glaze. Matt glazes containing more than 20% barium carbonate are unlikely to be food-safe. You can replace the barium with strontium carbonate but you only need three quarters of the amount to achieve the same matt effect with strontium (so using only 15% instead of 20% in the glaze recipe).

Colouring oxides dissolve in the glaze. At a certain level above 4% (for copper oxide), no more colouring oxide can dissolve and the remainder floats to the surface, taking on a black, metallic appearance (like the mixture of manganese and copper oxides brushed on rims to give a bronze effect, see metallic glaze section later in this book). This metallic surface is not food-safe. When you dissolve less than 2% colouring oxide in the glaze, it is likely to be food-safe, provided the glaze contains enough silica and alumina, and is fired to maturity. However, unstable, non-durable glazes are fine for decorative and sculptural pieces and this is mostly what special-effect glazes are used for.

Earthenware turquoise glossy, cone 04, 1060–1100°C (1940–2012°F).

This glaze has high boron to enable it to melt at a low temperature (Unity molecular formula: Na_2O 0.2, CaO 0.8, B_2O_3 1.0, Al_2O_3 0.32, SiO_2 3.0).

Calcium borate frit 39
Soda feldspar 27
Whiting 5
China clay 6
Flint 23
+
Copper oxide 1

Low-temperature earthenware glazes rely on frits to enable them to melt at around 1100°C (2012°F). Frits containing boron will make a more durable glaze than high-alkaline frits or lithium carbonate.

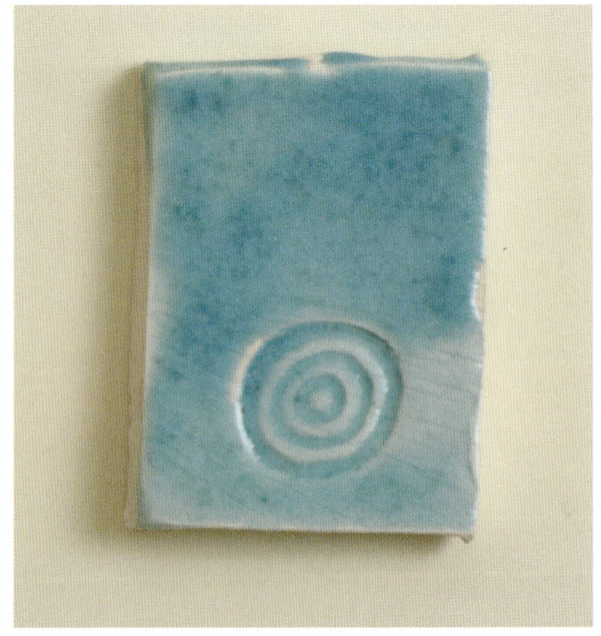

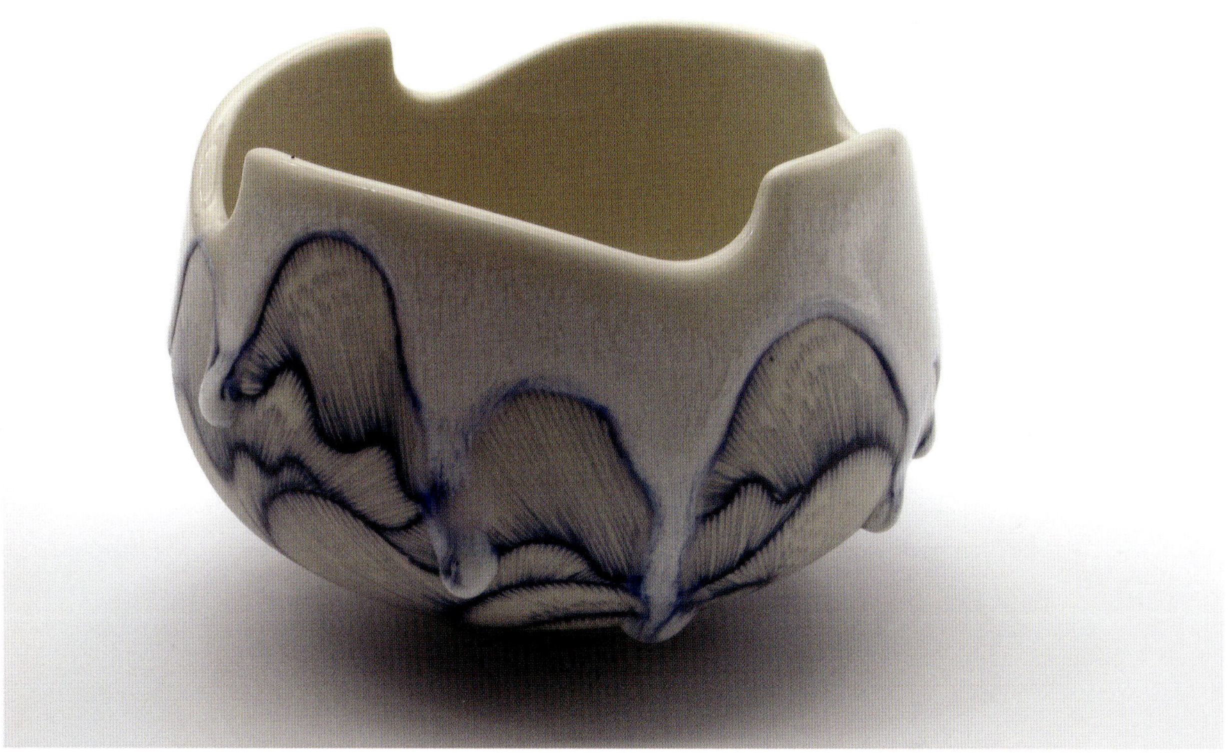

Noelle Hoover, Single-Tooth Productions, slipcast stoneware bowl, underglaze line drawing, drippy glaze, Indianapolis, USA.

The right firing temperature

Under firing results in weak, dry, rough glazes. If a glaze is not melted enough to form a glass, it will eventually crumble and break down when washed in the dishwasher and during daily wear when stacking or scraping with metal cutlery. You can tell if your glaze is under-fired as it will be a dry, unpleasant matt, not the shiny gloss or stony matt of a properly fired glaze. True matt glazes are those in which the glaze has melted and then recrystallised on cooling and can be stony or satin matt. On the other hand, over-fired glazes run a lot and may be blistered (although running glazes may well be what you want). In general, a durable glaze will have the maximum silica and alumina appropriate to a particular firing temperature (see glaze limits chart on p.46), but it is also important that the glaze has completely melted. Borax frit and calcium borate frit can help to melt glazes fired to temperatures from cone 8 (1260°C/2300°F) down to cone 04 (1060°C/1940°F). The lower the firing temperature, the more frit is added, with up to 90% frit for some earthenware and raku glazes. Some frits are almost a glaze on their own, but there is always some china clay, ball clay or bentonite in the glaze recipe to help suspend the frit in water.

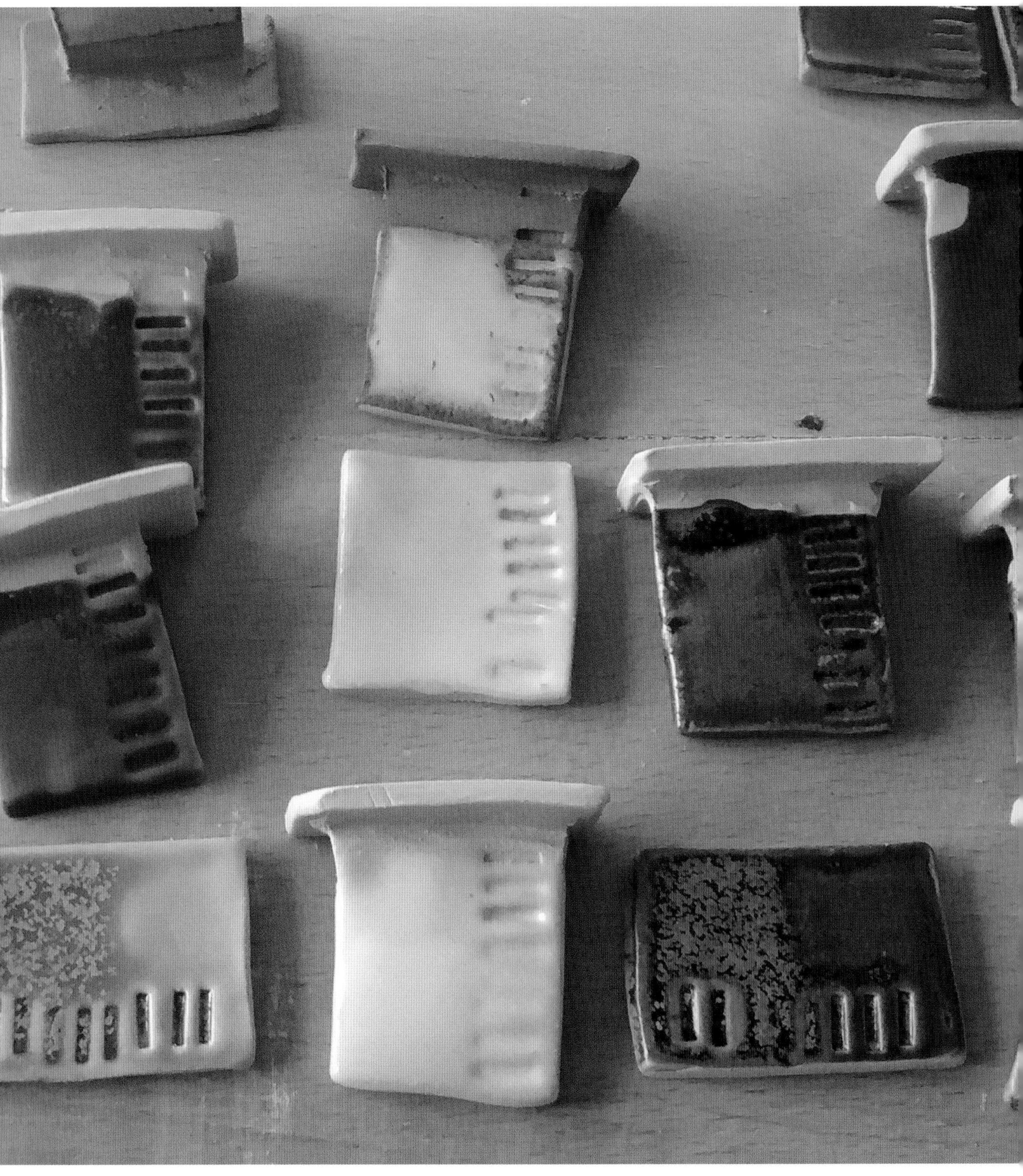

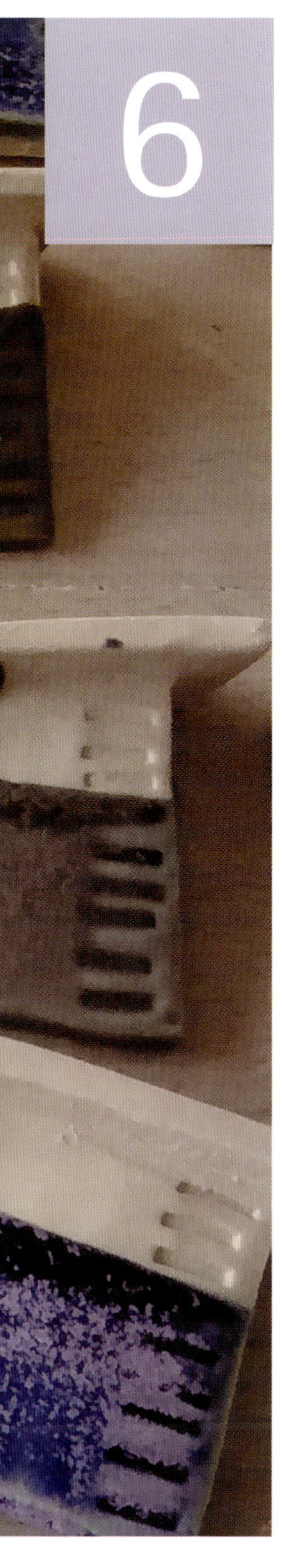

6

Creating and testing glazes

Glaze tests: Test tiles, blends and grids

Look for glaze recipes in books and online. Mix up and test small 100g batches of glaze. It is possible to adjust the texture, fluidity and glaze fit to the clay body by changing the amounts of silica and clay in the glaze. Low clay content may make the glaze runny, while a high clay content may cause the glaze to become matt. A lichen crawl texture can be provided by the addition of magnesium carbonate or zinc oxide and china clay. Once you have found a base glaze you like, whether glossy or matt, it is a good idea to test it with additions of various colouring oxides. An easy way is to paint lines of colouring oxide mixed with water onto a biscuit-fired tile and then dip into your base glaze and fire in the kiln. You will be able to see how the various colouring oxides react with your glaze. A more precise method is to weigh small quantities of colouring oxide and add them to 100g (dry weight) batches of glaze (see glaze tests that follow).

Glossy transparent glaze base, cone 9, 1260°C (2300°F).
Potash feldspar 34
Borax frit 14
Whiting 11
China clay 13
Quartz 23
Dolomite 5

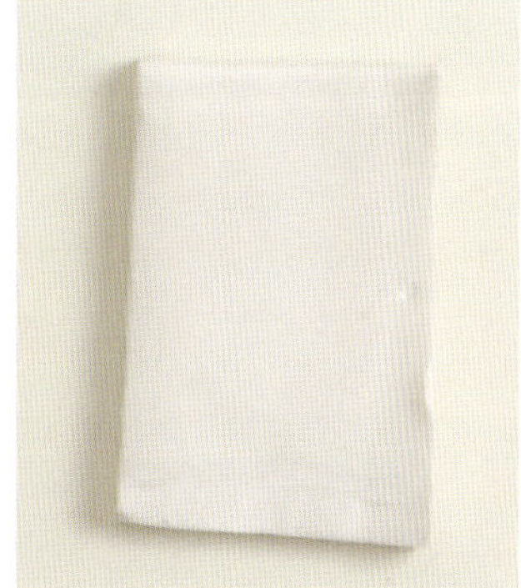

Glaze colour tests

Mix up a 1kg batch of base glaze (any glossy or matt glaze without colouring oxide additions). You can do this by multiplying the glaze recipe by 10 and weighing out each of the ingredients in grams. Then add the dry powder to about 750ml water and stir thoroughly to evenly distribute all the ingredients. Adjust the glaze thickness until it is somewhere between milk and single cream. If it is too thick, add more water. If it is too thin, leave to settle for a few hours and pour off water from the top. Divide all the glaze mixture equally into ten plastic cups (there will be more than 100g in each cup; 100g dry powder plus water). Then weigh out a small quantity of each colouring oxide and add a different oxide to each cup. You will need accurate scales, either digital or triple-beam balance.

LEFT: Magnesium crystalline matt test tiles with various colouring oxides (recipe on p.113)

Suggested quantities of colouring oxides to start with are:
0.5g cobalt oxide
1g copper oxide
0.5g chromium oxide
2g iron oxide
5g rutile
5g ilmenite
1g nickel oxide
2g manganese dioxide
5g tin oxide
5g zirconium silicate

Stir and sieve each coloured glaze separately through an 80 mesh sieve and dip a small test tile. You can dip half the tile again to get a double thickness of glaze, and even dip once again to get a triple thickness on part of the tile. Make sure the glazes are stirred well each time before dipping. Wipe excess glaze off the back of the tile and number on the back using an underglaze pencil or a mixture of iron and manganese oxides in water.

Test tiles should be made using your usual clay and making method. You can roll out a slab and cut rectangles, or throw a wide cylinder on the wheel and cut it into sections. Tiles can be placed flat on the kiln shelf or you can make L-shaped, vertically standing tiles which will show the fluidity of the glaze when fired. It is a good idea to add some texture to the tile by impressing a textured object or stamp. The tiles should be biscuit-fired to your usual biscuit temperature, or around 990°C (1814°F). Remember it is a good idea to use a test shelf in your kiln when test-firing new glazes.

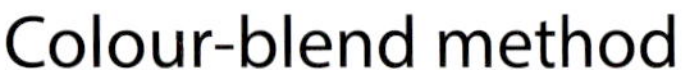

Colour-blend method

To make the most of the glaze colour tests above, each colour can be mixed 50:50 with every other colour, using a tablespoon or syringe. Each glaze must be mixed thoroughly before blending, then a tablespoon of each of two glazes is measured, mixed together and applied to a test tile by dipping or pouring.

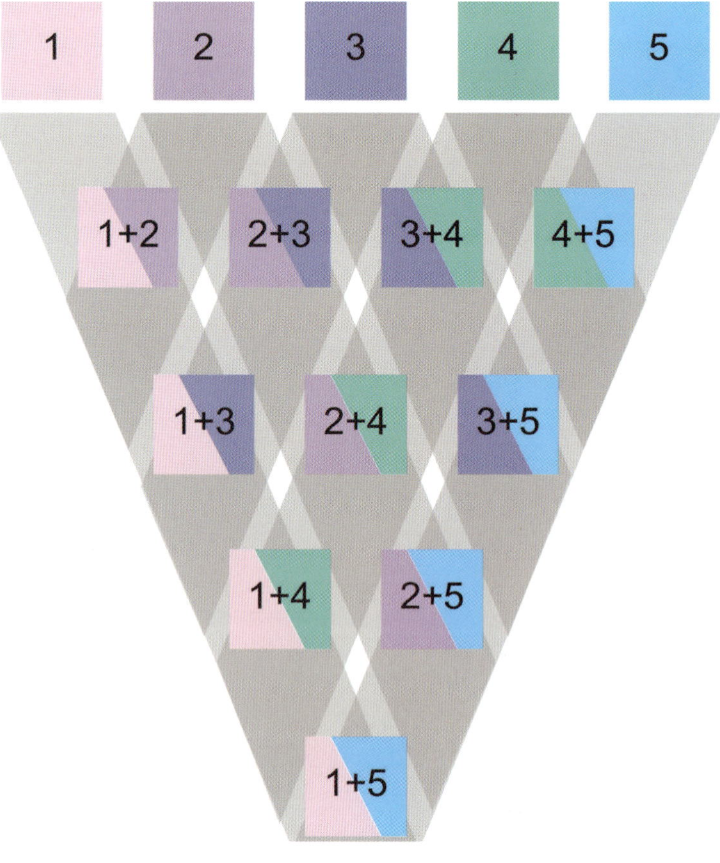

Using this wet blending method, the combined glazes do not need to be sieved again. Label the back of each test tile as shown in the diagram: for example, if glaze 1 is mixed with 2, label the test tile 1+2.

After the test tiles are dipped into each coloured glaze, fire them to your usual glaze firing temperature. The test tiles will show how the base glaze reacts with each colouring oxide. If there are particular results you like, you can do a series of more refined tests using the same oxides. This method is good for testing blends of two oxides; for three oxides you may need to do a triaxial blend (see p.56).

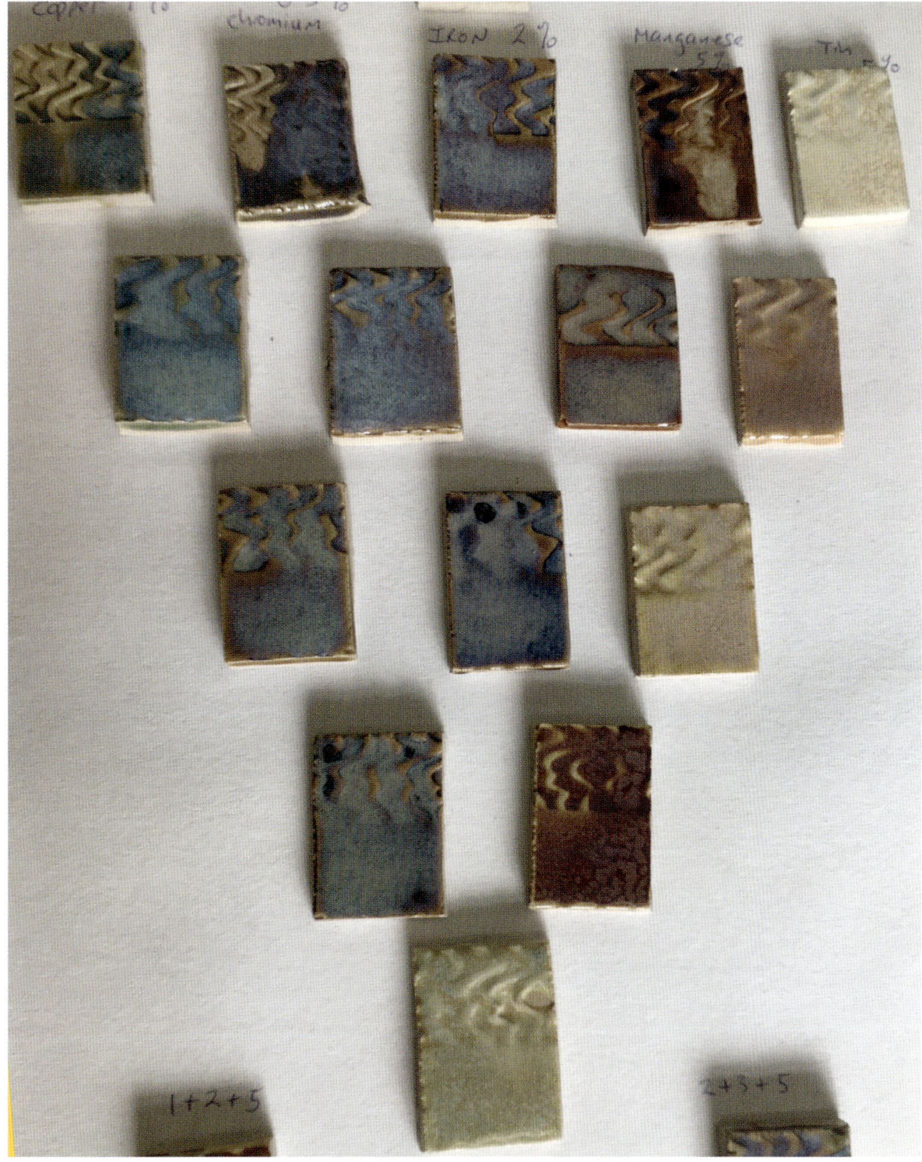

Glossy transparent glaze-base glaze from p.51 with 5% rutile and then tested with (first row left to right): 1% copper oxide, 0.5% chromium oxide, 2% iron oxide, 5% manganese dioxide and 5% tin oxide. All other tiles are 50:50 mixes with first row. This corresponds to the glaze blend diagram shown. Glaze tests made by Diane Brewster at Forest Row School of Ceramics.

LEFT: Diagram to show colour-blend test using five different oxides mixed in the same base glaze. Each one is mixed 50:50 with all the others, to create every possible combination of two oxides.

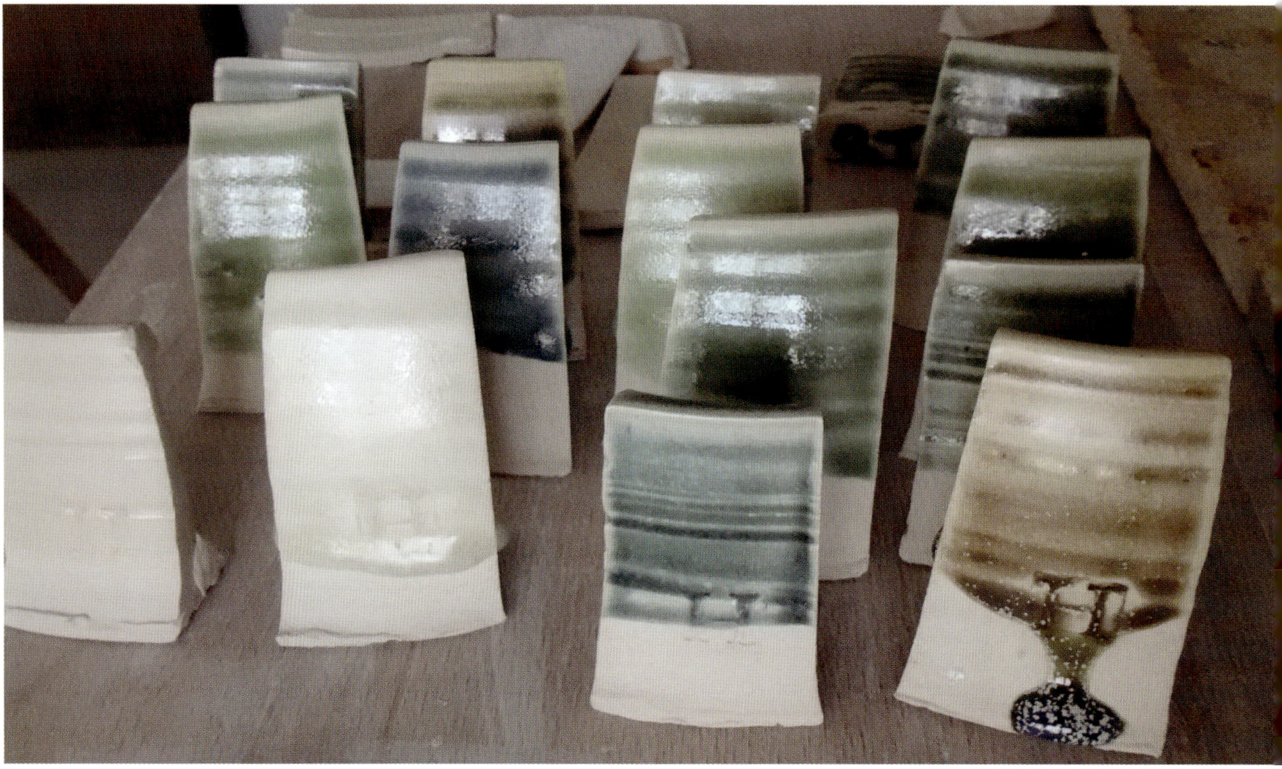

Test tiles with magnesium matt glaze given on p.113 with various oxides, fired to cone 9.

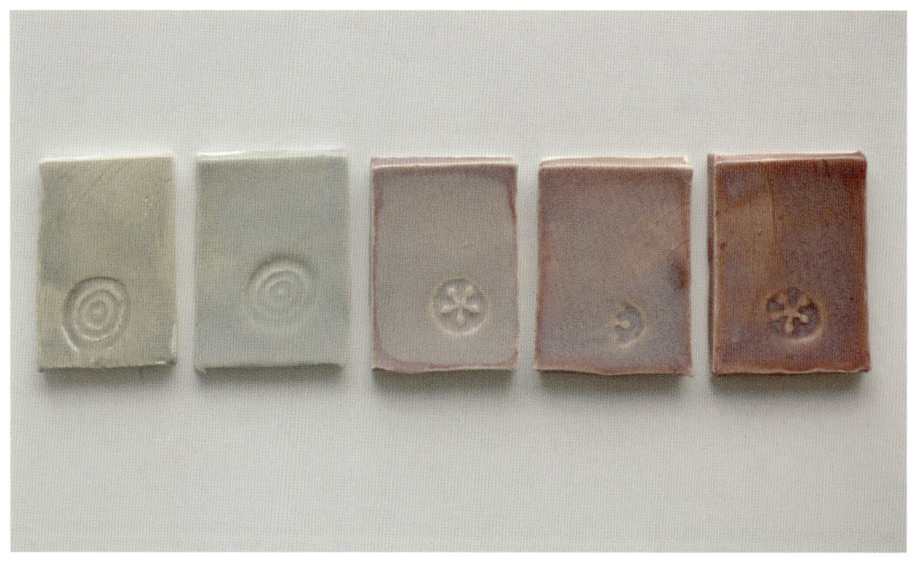

Line blend, left to right increasing whiting in a chrome-tin earthenware glaze.

Line blends

A line blend is a base glaze with increasing amounts of colourant added. For example, you could try adding copper oxide 0.5, 1, 1.5, 2, 2.5%, mixing, sieving and dipping a tile after each addition. I prefer to use small amounts of oxides to get pale, watery colours. If you use more than 5% copper oxide, the glaze will become saturated and will be a metallic black when fired.

Cross blends

You can also try mixing two oxides together such as cobalt and manganese, copper and rutile, chromium and tin, cobalt and nickel, copper and tin. An easy way to do this is to mix together two of the cups of coloured glaze made in the colour-blend tests on p.52. You can also mix smaller volumes such as a tablespoon or 25ml measured using a plastic syringe. Make sure the glazes are stirred well before measuring in this way.

A cross blend is a series of tests using two colouring oxides, where one oxide increases as the other decreases, for example:

Copper oxide	0.5	1	1.5	2	2.5 %
Rutile	10	8	6	4	2 %

You can do this by weighing out the colouring oxides individually or more simply by wet blending, adding tablespoons of each of the copper and rutile glaze tests made previously; see glaze colour blend tests on p.52. Mixing 1g copper and 5g rutile in 200g of base glaze would result in a glaze with 0.5% copper and 2.5% rutile. You would need to mix at least 500g dry weight of glaze to have enough volume to complete the cross blend above.

Colouring oxides are best for making blues, greens and browns. Grey and black can be made using combinations of two or more oxides. Cobalt and nickel oxides will combine to make grey. You can also make interesting greys, blues and blacks using three colouring oxides: cobalt, iron and manganese or cobalt, nickel and manganese. Glazes made using colouring oxides will be transparent and will have more depth than those made using commercial stains, which tend to be very solid and opaque.

Line blends and cross blends are also useful methods for testing commercial stains such as yellow and red. Rather than buying an orange stain, you can create your own varying shades of orange by mixing yellow and red in different proportions. Stains are generally weaker colourants than oxides, and 5–10% stain is needed to colour a glaze.

Triaxial blends

In a triaxial blend, three coloured glazes are mixed together in varying proportions. You only need to mix the three corner glazes of the triangle; the others can be mixed by wet blending, using a tablespoon or syringe. Mix up at least 500g of each of the corner glazes, sieve each glaze, place in a container and lay out the three containers in a triangle. With a tablespoon or syringe, take a proportion of each glaze, e.g. 20:80, 40:60, place in a cup and mix together, then pour on to or dip a test tile.

After firing the glaze test tiles, lay them out in a triangle and choose which you prefer. To calculate the amounts of colouring oxides in a particular test tile, multiply the percentage of oxides by the proportions of glaze in the test.

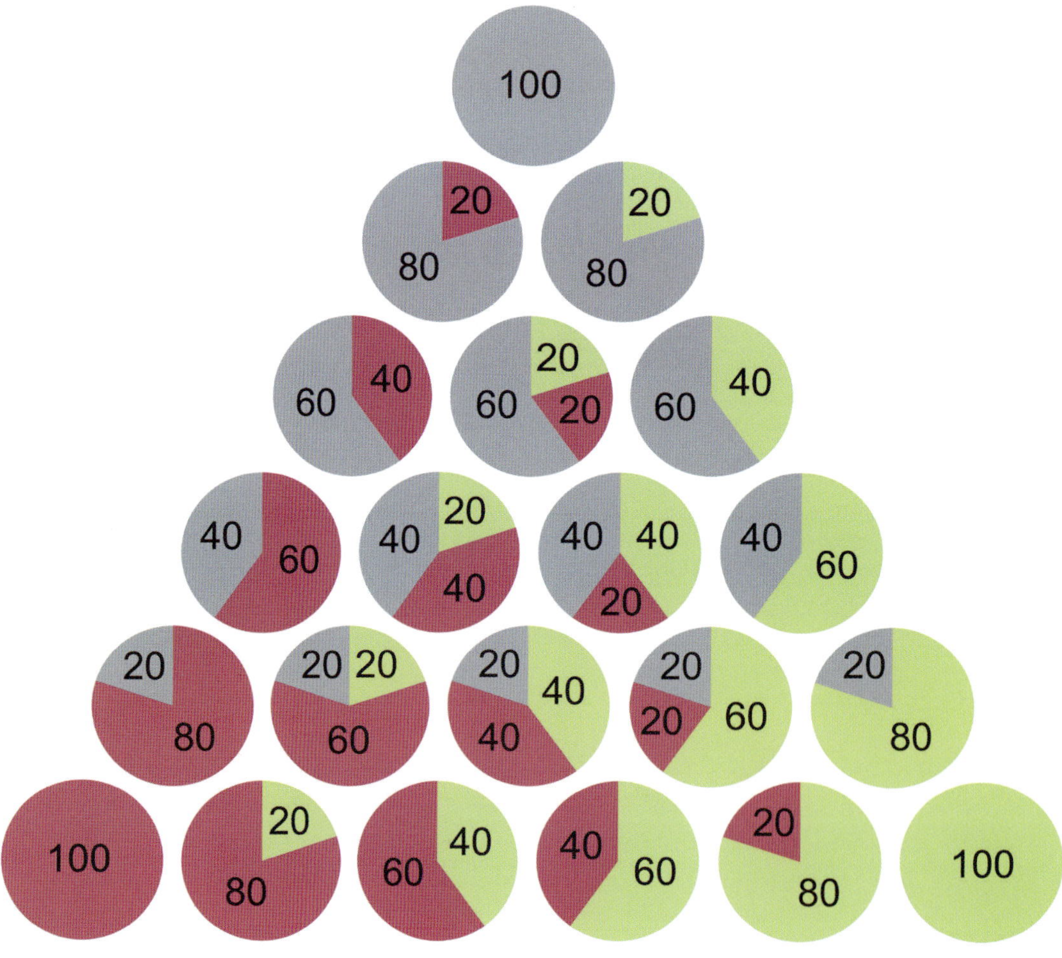

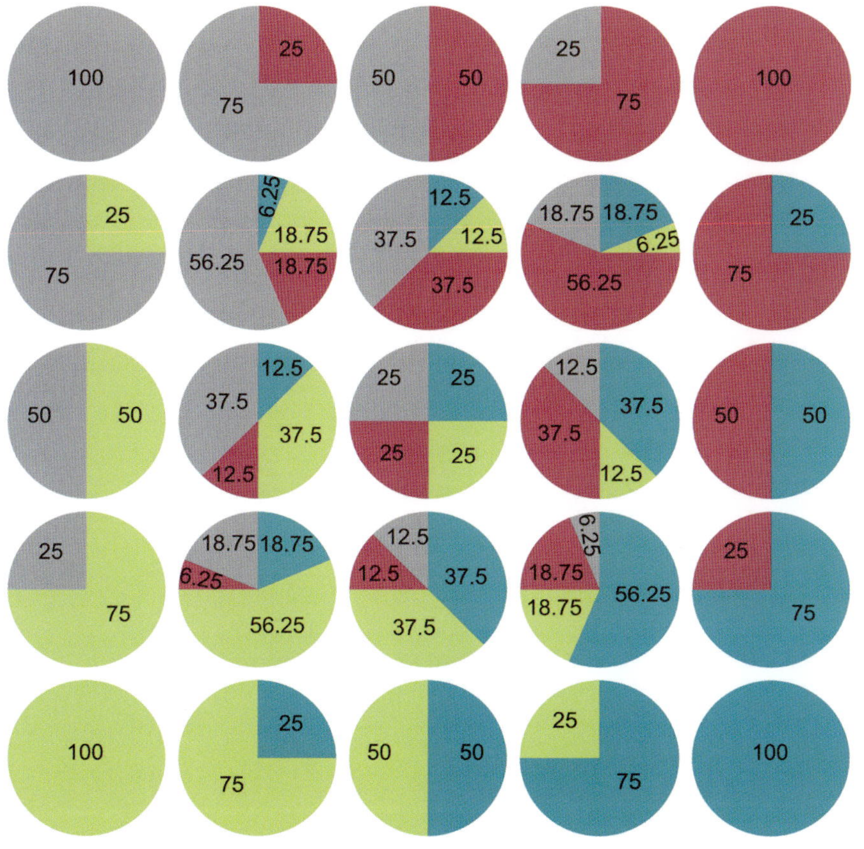

Quadraxial blends

In a quadraxial blend, you can combine four materials to make a glaze, such as feldspar, wood ash, clay and quartz, or four different coloured glazes to make colour blends. You only need to mix up and sieve the four corner glazes and the rest can be made by volumetric blending of the four corner glazes in the varying proportions shown. This is best done with a syringe.

Currie grids

A Currie grid, invented by Australian potter Ian Currie, is a 7x5 biaxial grid where the fluxes and colourants remain constant but the clay and silica are varied. The tests are made from four corner glazes: 1) the base fluxes with no clay or added silica, 2) with high clay (40% kaolin), 3) with high silica (50% silica), and 4) with both high clay and high silica (25% kaolin, 40% silica). These four corner glazes are mixed with water and sieved. The rest of the grid is generated by combining the four corner glazes in varying proportions, measuring wet volume using a syringe (see ian.currie.to and grid on p.85).

Results and records

Always make a note of the glaze recipe, test tile number and fired result. It is important to record your research as you may want to come back to a particular glaze some time later. Keep your glaze tests and notebooks in a safe place in the studio. Some potters make holes in their test tiles and hang them on a board. You can also make test tiles to attach to the glaze bucket.

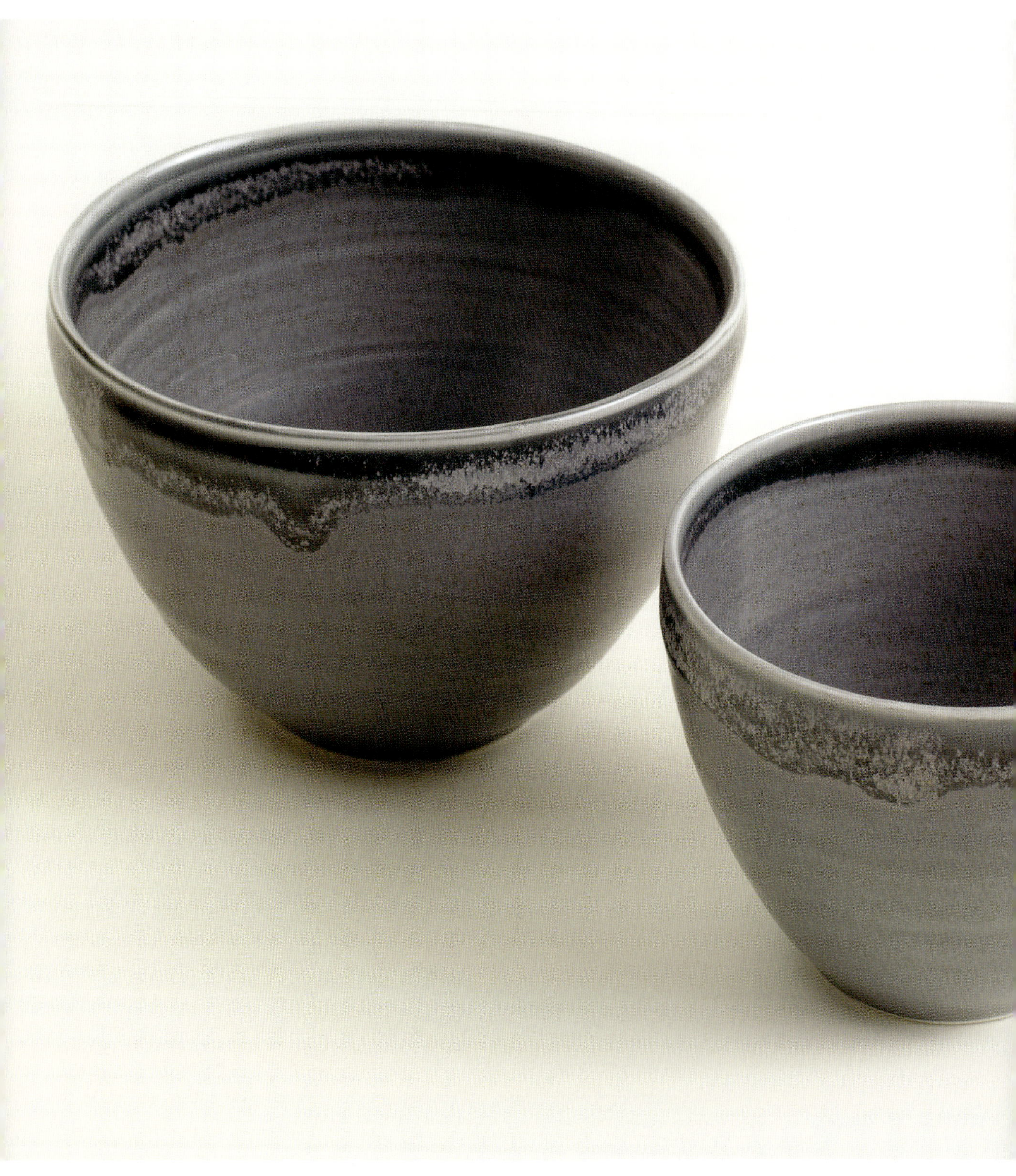

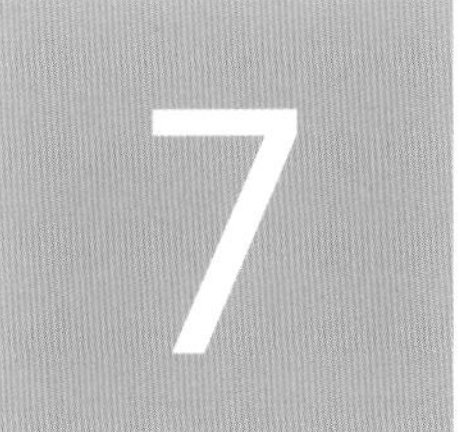

Glaze mixing and application

Once you have tested a few glaze recipes and found one that gives the fired result you want, you will need to mix up a larger batch of glaze so that you can start applying it to your pots. The simplest application methods are pouring, dipping and brushing. Bubble glazes need a special approach however, see p.62 for this method.

In order to dip the pots in the glaze, you will need to mix up at least 5 to 10 litres. I usually mix 7.5 kg of glaze in a 10 litre bucket with a lid. The glaze recipes shown are multiplied by 75 to give the batch weight in grams. Start by half-filling the bucket with water, then carefully weigh out the glaze ingredients and tick them off the list one by one. It is best to add the china clay or ball clay first, as this helps to suspend the heavier ingredients in the water. If you add the feldspar first, it is likely to settle in a hard mass at the bottom of the bucket. Then leave all the ingredients to slake in the water for a few hours before mixing and sieving three times through an 80 mesh sieve. You may need to add water to help flush the glaze through the sieve. If it ends up too thin, you will need to leave to settle overnight and remove water from the top, so if possible, make the glaze the day before you need to use it. The consistency should be between milk and single cream, but some glazes such as ash glazes need to be thicker. I stir by hand from the bottom of the bucket, and if it is the correct thickness, I can just see the skin on the back of my hand through the glaze coating. However, ash glazes are caustic, so always wear rubber gloves when mixing these. When the consistency is right, mark the level on the glaze bucket and use the same level every time you make a new batch of that glaze. Alternatively, you can weigh 100ml glaze and record the weight, then make sure the weight is the same each time you mix a new batch. The weight of glaze in grams divided by the volume in millilitres is known as the specific gravity, so if 100ml glaze weighs 140g, its specific gravity is 1.4. If you measure the specific gravity, your glaze consistency and application thickness will be consistent from batch to batch. Some potters prefer to use a hydrometer to measure the specific gravity but I prefer to calculate by weight. In general, glazes need to be applied more thickly on stoneware than on porcelain. Most special-effect glazes also need a thick application, including crackle, crawl, volcanic and oil-spot glazes. The glaze thickness on the pot can be anywhere between 1 and 3mm.

The glaze consistency needs to be maintained over time and water added if the glaze has thickened. If the glaze seems too thin a few weeks after it has been mixed up, it may need to be flocculated. Some glaze materials such as nepheline syenite leach sodium into the water, turning the water alkaline. This causes the glaze to

Bowls with runny glossy glaze applied on rims on top of a dark grey matt glaze. Thrown porcelain by Linda Bloomfield.

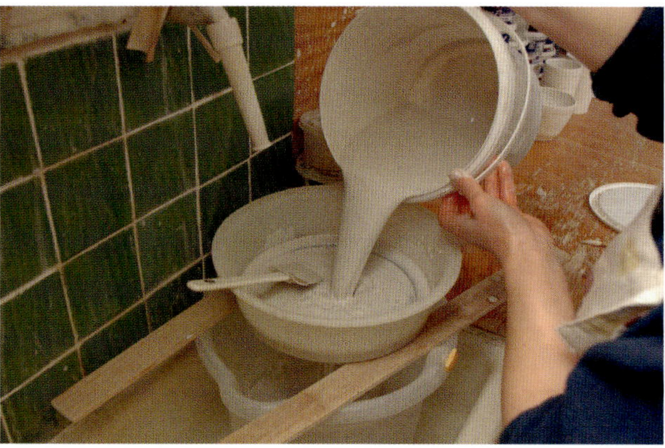

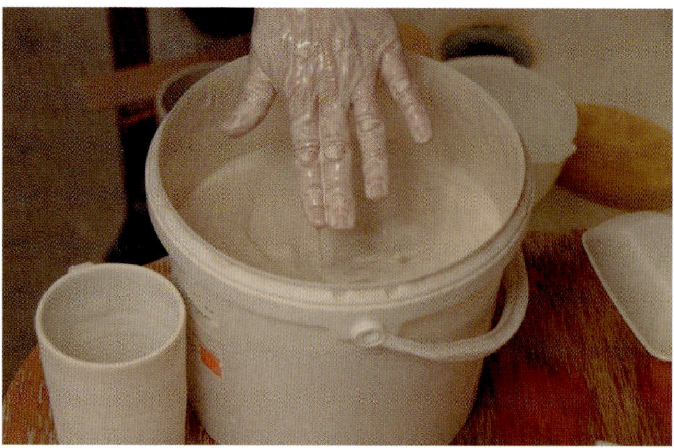

ABOVE LEFT: Sieving the glaze.

ABOVE RIGHT: Checking the thickness of the glaze. This particular runny porcelain glaze needs to be applied quite thinly.

settle in a hard layer on the bottom. Adding Epsom salts, just a few drops of a solution made from a teaspoon of salts dissolved in warm water, can correct this problem. A small amount of bentonite mixed with water will also help. However, if the glaze water becomes acidic, which can happen in bone ash glazes, the glaze may crack when it dries on the pot. In this case, add a drop of deflocculant such as sodium silicate or Dispex.

Preparation for glazing

You will need biscuit-fired pots, fired to around 990°C (1814°F) for stoneware or 1060°C (1940°F) for earthenware. Some potters raw-glaze unfired pots, but biscuit firing allows you to wash off the glaze, leave to dry and start again if anything goes wrong. If you apply glaze straight after the biscuit firing, it should not be necessary to wash the pots first. If, however, they have been on a shelf for some time in the studio, any dust should be sponged off before applying glaze. Make sure you leave time for the pots to dry; glazing onto damp biscuit ware can sometimes cause pinholes.

Pouring

Pouring is the best application method for the inside of pots, or if you need to glaze a pot which is too large to dip. The glaze should be stirred very thoroughly, and stirred again between glazing pots, otherwise it may start to settle in the bucket. Scoop up a cup or ladle full of glaze and pour carefully into your pot, swill around for a few seconds to coat evenly and pour out. You may need to use a funnel to glaze the inside of narrow-necked bottles. If you want a different glaze on the outside, wipe any glaze drips from the outside using a sponge and leave to dry. Very thin-walled pots will also need to be glazed on the inside first, then left to dry before being glazed on the outside. Very large pots can be placed on a whirler or potter's wheel and rotated slowly while glaze is poured over the outside.

Dipping

Dipping is a good way to apply glazes when you need a very even application. When dipping in volcanic glazes, make sure the glaze is thoroughly stirred so that the silicon carbide particles have not sunk to the bottom. Hold the pot sideways by the rim and foot, between finger and thumb, or using glazing tongs. Dip it swiftly into the glaze and out again, pouring out all the glaze from the inside of the pot. If you are only glazing the outside, hold the pot by pressing your fingers against the inner sides of the pot and dip up to just below the rim. Place on a clean surface and don't touch the pot again until the glaze has dried. You can finish by dipping the rim upside down or brushing glaze onto the rim. Any drips should be scraped flat with a sharp knife. Any finger marks or bare patches can be touched up using a brush.

BELOW LEFT: Dipping a mug.

BELOW RIGHT: Sponging the bottom of a mug. The glaze was first scraped off using an old credit card.

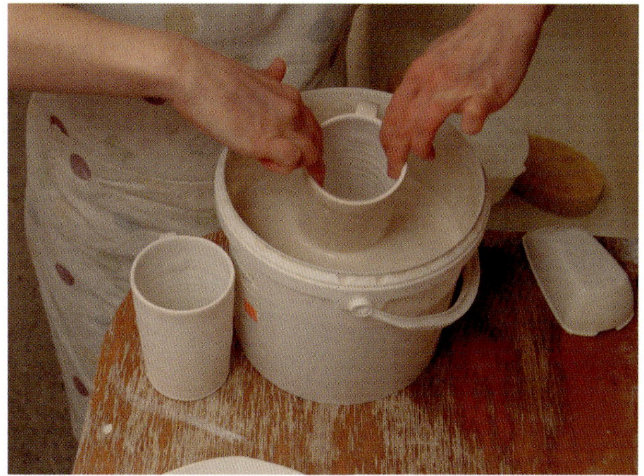

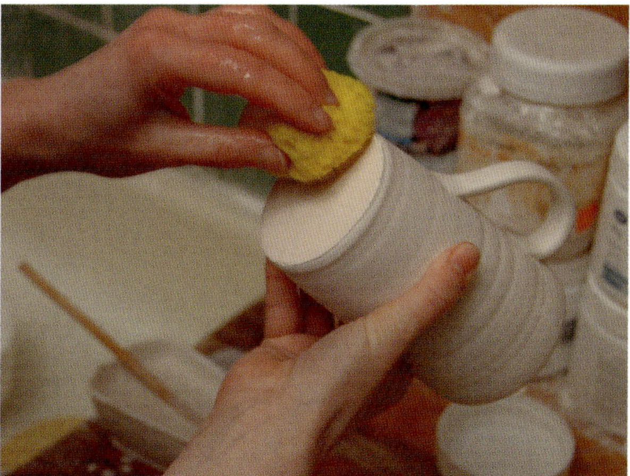

Brushing

To make the glaze easier to brush on, you can add CMC (carboxy methyl cellulose) or gum arabic to the glaze. Two or three coats of glaze will be needed to get an even coverage. I usually apply brush strokes round the pot, then go round a second time, starting each brush stroke in between the previous brush strokes. You can also apply brush strokes in a perpendicular direction to the previous coat. Potters often use Chinese calligraphy brushes or wide Japanese hake brushes for covering larger areas.

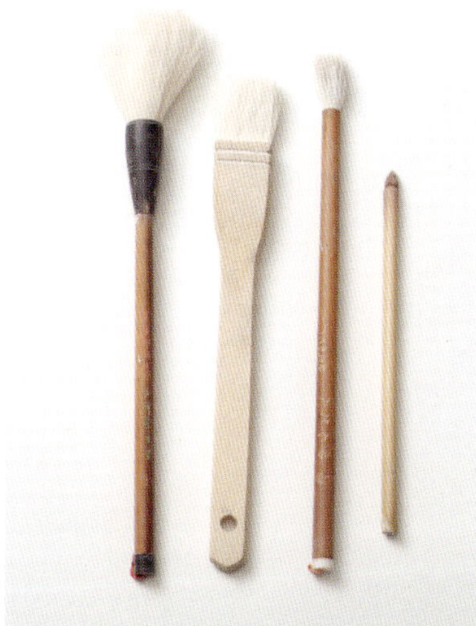

Brushes for glazing.

You can mask areas you don't want glazed by applying wax resist, latex or masking tape. However, I find it easier to scrape glaze off the base using an old credit card and then carefully sponge the base. When throwing pots on the wheel, I always bevel the edge of the base to make a neat line to sponge back to. This also helps to prevent the glaze from sticking onto the kiln shelf.

Spraying

The advantage of spraying is that you don't need to mix up a large batch of glaze. The drawback is that you need a lot of equipment: a spray booth, spray gun and compressor. A respirator mask should also be worn while spray glazing. The pot is placed on a whirler and rotated while spraying glaze on the inside and outside in several thin coats.

Bubble glaze application

Bubble glazes are made using a mixture of underglaze, water and a small amount of detergent. By blowing steadily through a straw into the underglaze mixture, bubbles of various sizes are produced. The bubbles can be applied to bone-dry or biscuit-fired pots. The pot is rotated into the mass of bubbles until the surface is covered. For plates, the bubbles can be poured directly onto the flat surface. In this case the underglaze mixture needs to be relatively thick, like cream. The underglaze can then be covered with a clear glaze and fired.

Emma Allington making bubble glaze using black underglaze mixed with water and detergent. Emma applies the bubbles to bone-dry clay, then glazes the inside after biscuit firing and polishes the unglazed surfaces after glaze firing.

Emma Allington, *Collins, With A Twist,* bubble-glazed porcelain collection inspired by the shapes of cocktail glasses.

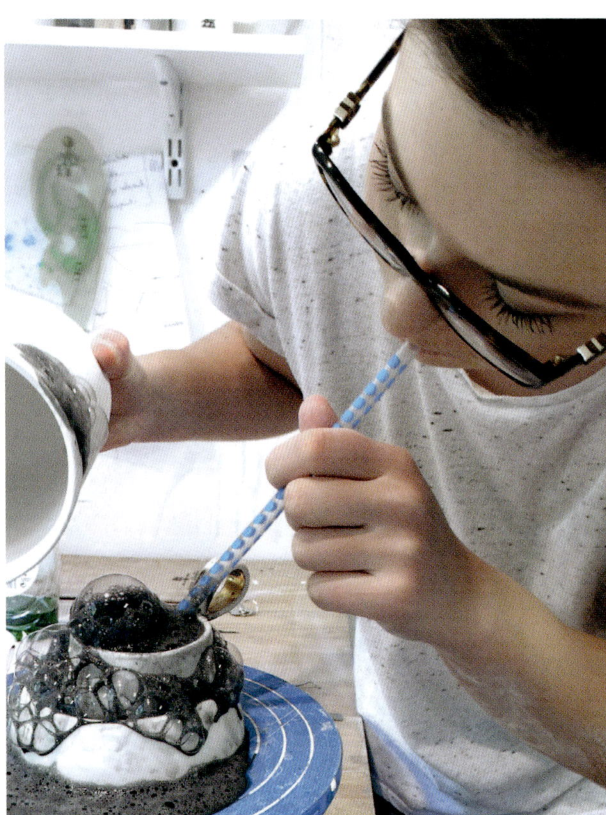

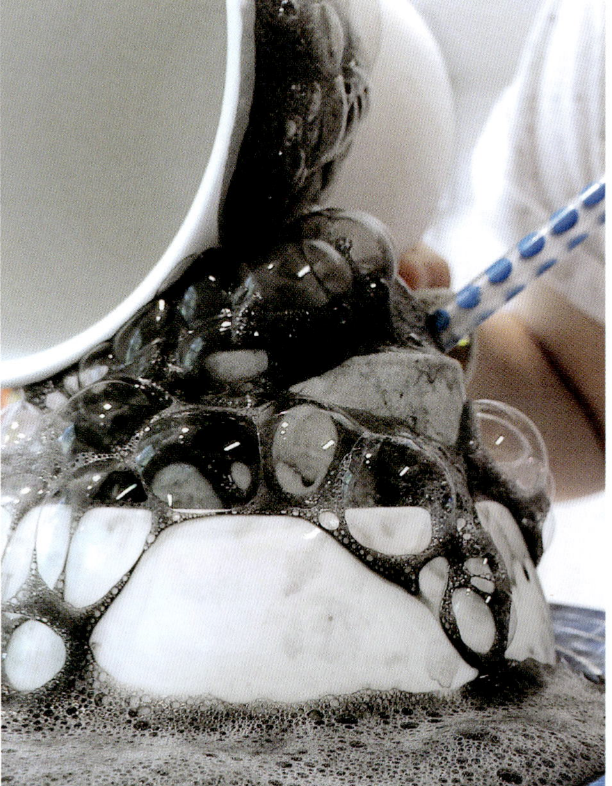

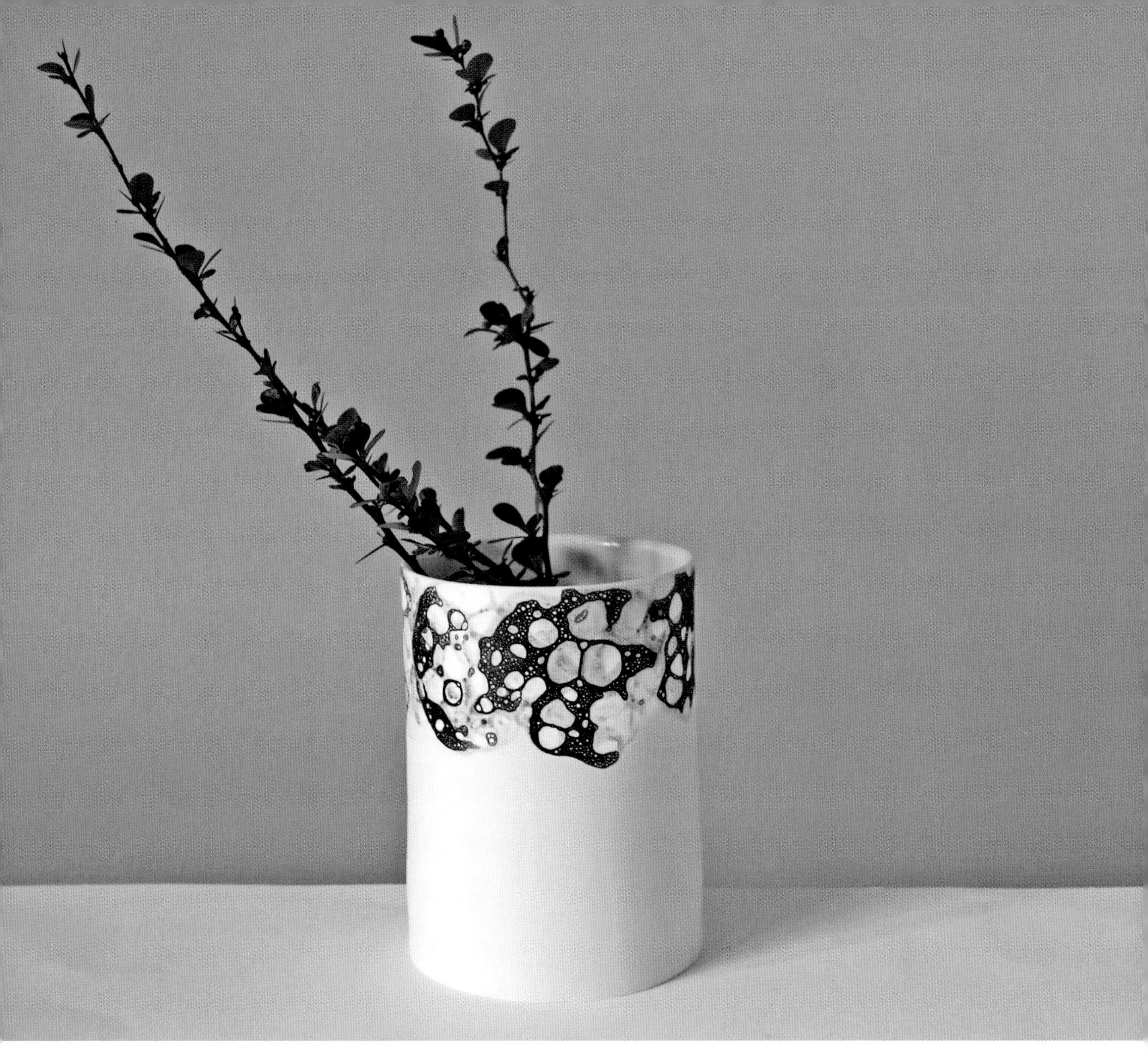

Layering glazes

Underglaze colours or oxides can be brushed on before applying the glaze or a layer of slip containing colouring oxides. Rich effects can be obtained by applying several layers of different glazes, for example a white glaze over a darker glaze or a runny glaze over a matt glaze. The glazes often react, producing interesting effects. Try out combinations of various glazes. The effect will be different depending on which glaze is underneath and which is on top. The combined glaze layer may become very thick, so make sure you test in the kiln to see if it runs before applying to a pot. Always use a test shelf or make a clay tray to test- fire on. If you want a really thick layer of glaze it may be necessary to apply one coat, biscuit-fire to harden, then apply another coat.

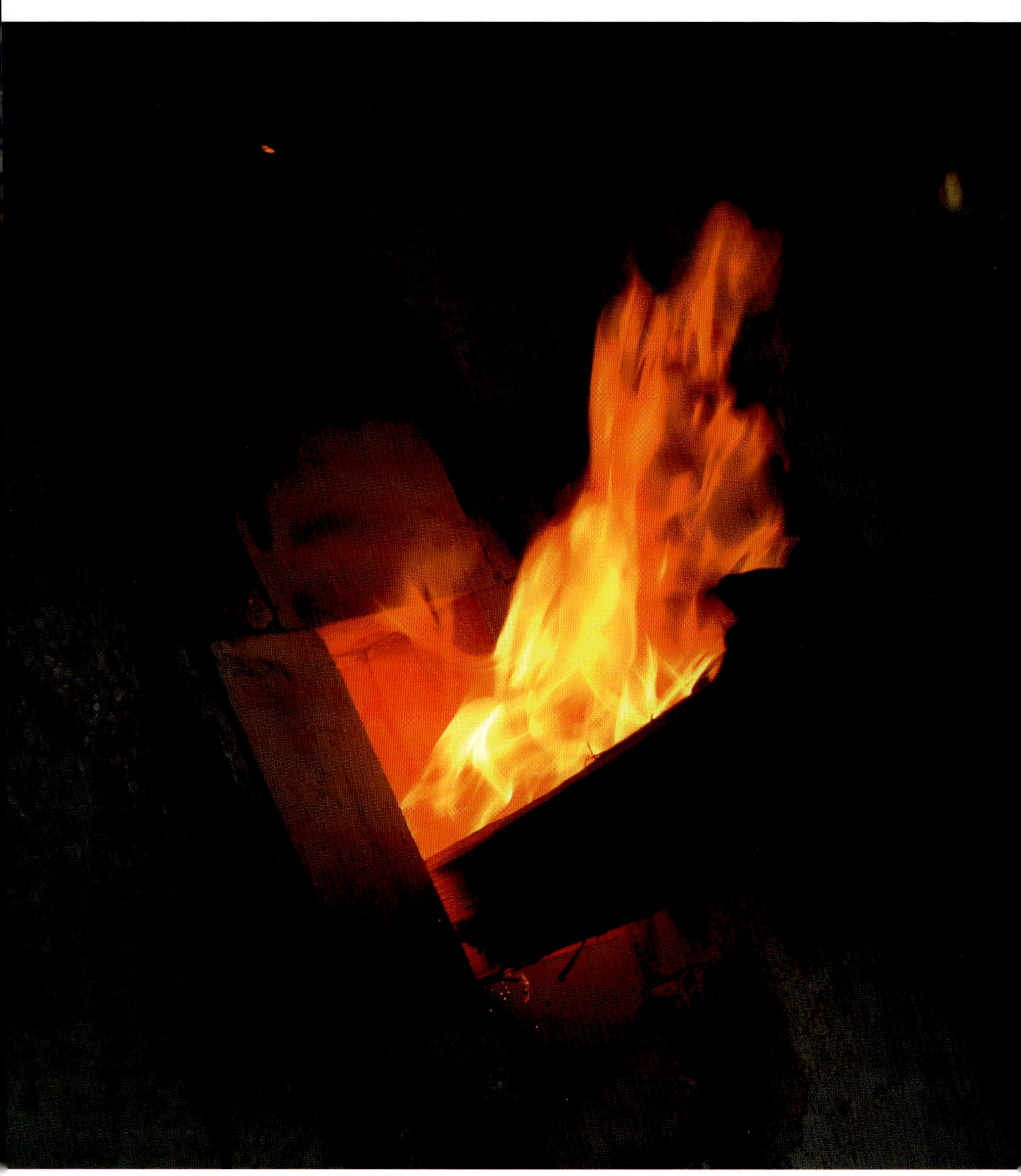

8

Firing

Pots are usually fired twice. The two firings are known as the biscuit (or bisque) firing and the glaze firing. After making the pots, they should be left until they are bone dry before firing. The first firing, known as the biscuit firing, should proceed slowly to give time for any residual water to evaporate. If the pots are damp, it is a good idea to heat them to 80°C (176°F) and hold for several hours until no more water vapour emerges from the spy hole in the kiln. The firing can then proceed slowly (60°C/140°F per hour) until 550°C (1022°F), when all the chemically combined water in the clay has been driven off. At 573°C (1063°F), the crystalline quartz in the clay body increases in volume by 1%. This may cause cracking if the temperature in the kiln is increased too rapidly. After this point, the rate of heating can be increased to 100°C (180°F) per hour. Organic matter, sulphur and fluorine in the clay are burned off at 700–900°C (1292–1652°F). Biscuit firings usually go up to a temperature of around 1000°C (1832°F). The biscuit ware is then strong enough to withstand handling and glazing.

After glazing, the pots are placed back in the kiln, making sure that all the bottoms have been wiped clean of glaze and that no pots touch each other. The firing temperature should be increased slowly at first so that the water in the glaze can be driven off without forming steam and blowing off the glaze layer. At 500°C (932°F) the chemically combined water from glaze materials, such as clay and light magnesium carbonate, is driven off. At 600-800°C (1112–1472°F), carbon dioxide is given off from carbonates in the glaze, such as dolomite and whiting. Above 900°C (1652°F), the glaze starts to melt and bond with the clay body. Above 1000°C (1832°F), silicon carbide breaks down to form silica and carbon dioxide, which can form bubbles and craters in the glaze. Between 1025°C (1877°F) and 1232°C (2250°F), oxygen is released from copper oxide, manganese and iron oxide. The oxygen gas bubbles up through the glaze and can give oil-spot effects in certain glazes, depending on their melting temperature. During firing, the oxides in the glaze and clay body need time to melt and interact. It is a good idea to hold at the top temperature for around 30 minutes to allow the glazes to mature fully.

After the firing, the kiln can be left to cool naturally, or for crystalline glazes it can be held at 1060°C (1940°F) for several hours to grow the crystals. At 573°C (1063°F) and 226°C (439°F), quartz and cristobalite (two different phases of silica) in the clay body contract and undergo inversion, causing stress and cracking if draughts are let into the kiln. The kiln should not be opened until it cools to below 100°C (212°F).

Firing temperature is usually measured using a thermocouple inside the kiln. However, these are not always accurate and it may be better to use pyrometric cones to measure the heatwork in the kiln. Three cones are stuck in a pad of clay at a slight

Firing an anagama kiln. Volatile sodium salts and ash are carried by the flames through the kiln and come to rest on the clay surfaces, giving a toasted orange effect and glassy drips of melted ash. (See also chapter on ash glazes.)

angle. The middle cone is the target temperature, the left cone bends at a slightly lower temperature and the right-hand cone is for a higher temperature, to guard against over-firing. When the middle cone is fired correctly, it should bend until the tip touches the base. The cones can be viewed through the spy holes during firing (this requires careful positioning during packing), or after the kiln has cooled down, to record the temperature reached during a particular firing. The bending of the cones depends on time as well as temperature, as a longer time at a lower temperature will give the same heatwork as a shorter time at a higher temperature. Stoneware glaze firings can be to cone 10 (1280°C), cone 8 (1260°C) or cone 6 (1240°C). Some potters prefer to use lower temperatures to prolong the life of the elements in their electric kiln.

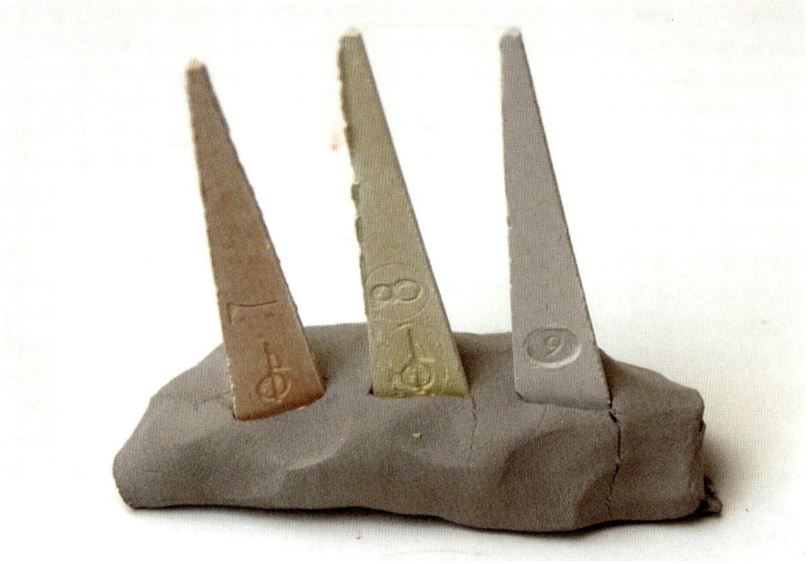

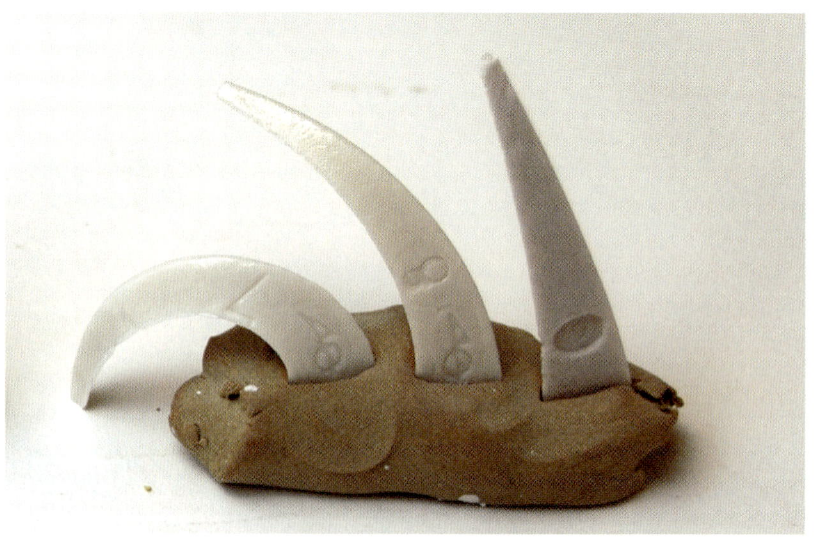

LEFT: Cones shown before and after firing. In this cone pack, only the left- hand cone 7 is fired to maturity with tip touching the base.

RIGHT: John Butler side-stoking his anagama kiln.

Oxidation and reduction

When pots are fired in an electric kiln, there is sufficient oxygen available for all the oxides in the clay and glazes to remain oxidised, so that, for example, iron oxide turns a tan yellow or brown colour. In fuel-burning kilns, the air supply can be restricted so that there is insufficient oxygen for combustion of the fuel, and a reduction state develops, where oxygen is drawn instead from the clay body and glazes. Iron oxide in the clay body and glazes is reduced to black iron oxide, promoting green celadon and black tenmoku glazes. The clay body turns grey and there is more interaction between the body and glaze, with iron spots bleeding through to the glaze surface.

Some special-effect glazes such as volcanic and oil-spot glazes need to be fired in oxidation. However, in an electric kiln it is also possible to produce local reduction effects by adding a reducing agent such as fine silicon carbide to the glazes. Using this method, only the oxides in the glaze will be reduced and the clay body will still be oxidised.

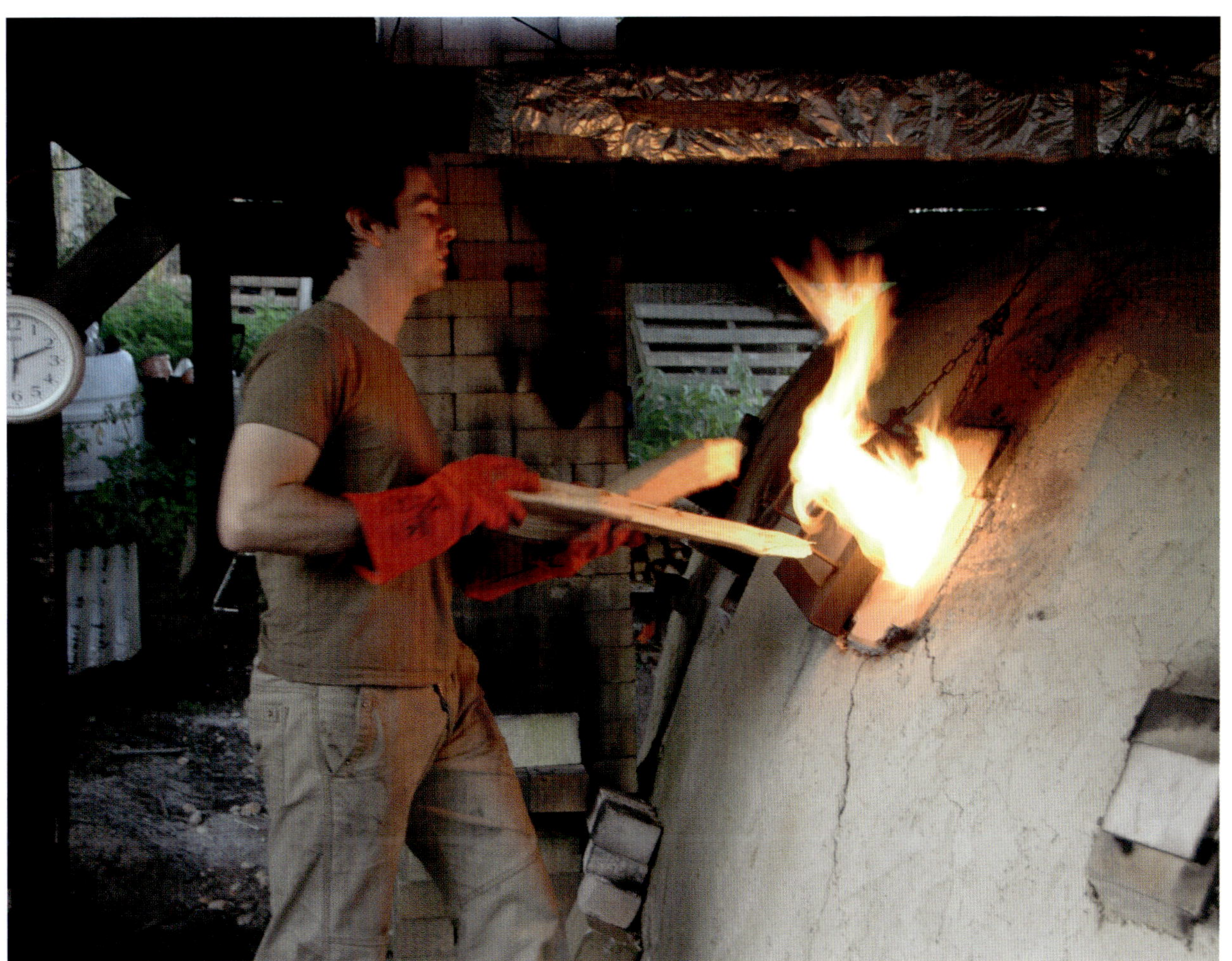

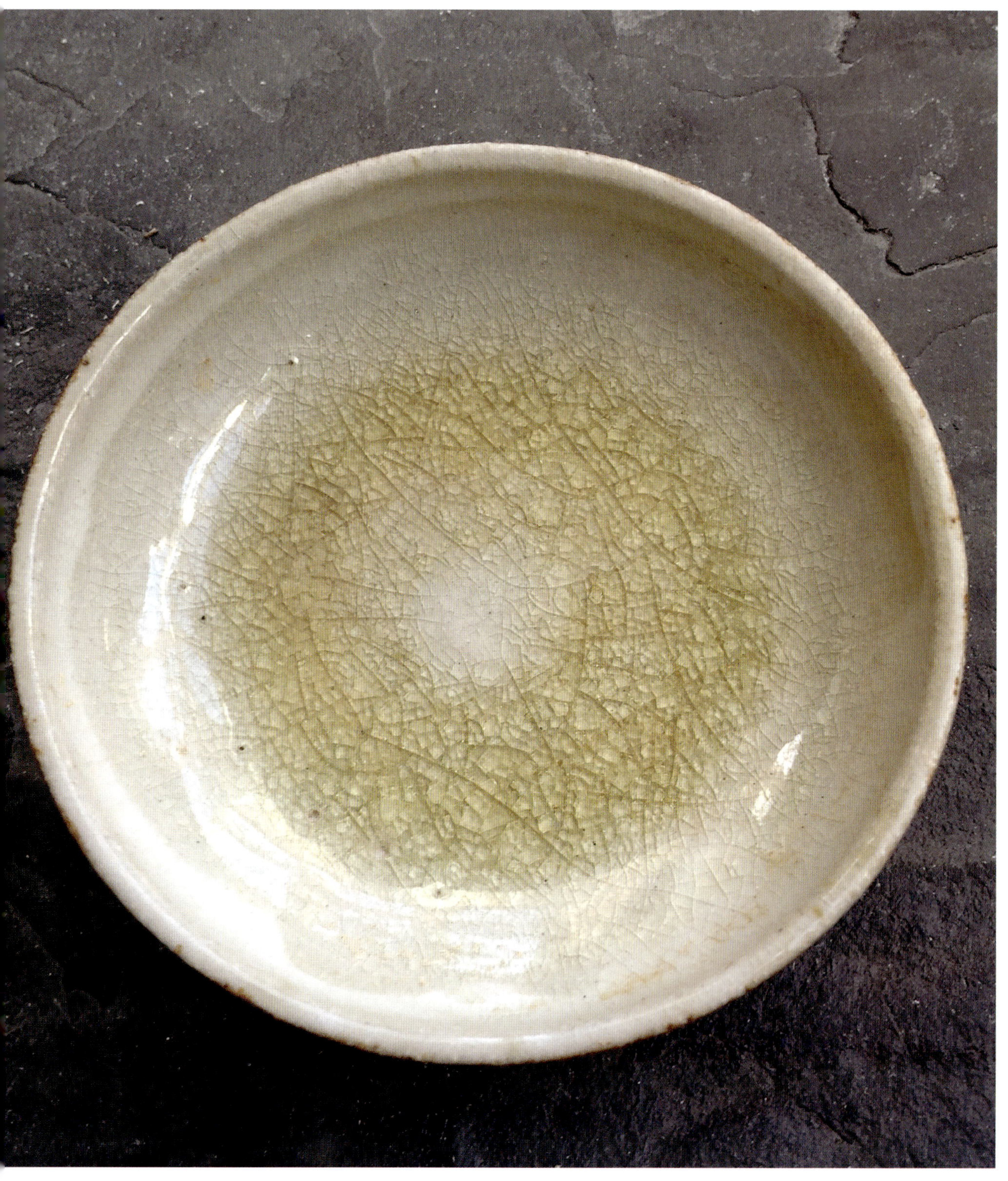

9

Glaze 'defects'

What is seen as a defect by one potter will be another potter's intended, carefully perfected technique and this is where the worlds of 'normal' and special-effect glazes meet. For example, if the glaze contracts more than the clay body during cooling, then cracks will appear in the glaze when the pot is removed from the kiln, accompanied by pinging sounds. This is known as crazing. It is not desirable on functional ware but some potters use it as a decorative effect, where it is called crackle (more on this in the next section). To correct crazing, a low-expansion material such as silica or talc (magnesium silicate) can be added to the glaze; conversely, to encourage crazing, a high-expansion material such as nepheline syenite can be added.

The opposite of crazing is called shivering, and is occasionally used as a special effect. The glaze contracts less than the clay body during cooling and starts to flake off rims and edges of handles, or can even crack the pot in two. This is a serious fault, which can be corrected by adding materials with high expansion, containing sodium and potassium, including nepheline syenite and high-alkaline frit. Reducing the silica in the glaze will also help.

Crawling is a defect sometimes used as a special effect which can occur when the glaze is applied too thickly and cracks on drying. It can be corrected by applying the glaze more thinly or reducing materials with high drying shrinkage such as clay and zinc oxide. Materials with high surface tension when melted can cause the glaze to form beads with bare patches in between. These materials include zirconium and tin oxide, which are used to opacify the glaze, making it white. Crawling is sometimes caused by greasy or dusty biscuit ware. If you think this is the case, the ware should be sponged with water and left to dry before glazing.

Other problems are caused by under-firing or over-firing the glaze. Underfiring can cause pinholes, while overfiring can cause blisters in the glaze. Pinholes are also caused by applying glaze onto biscuit ware that is damp or that has been fired too high, making it less porous. Reducing the biscuit-firing temperature can help. Soaking or holding the peak temperature at the end of glaze firing can heal pinholes. Gently rubbing the dried glaze over the holes before firing can also resolve pinholing. Blisters can be ground down and refired. Crawling, pinholes and blisters can be the result of too thick an application of glaze.

Linda Bloomfield, stoneware dish with crazed ash glaze, fired to cone 9 (1280°C/2336°F) in an electric kiln.

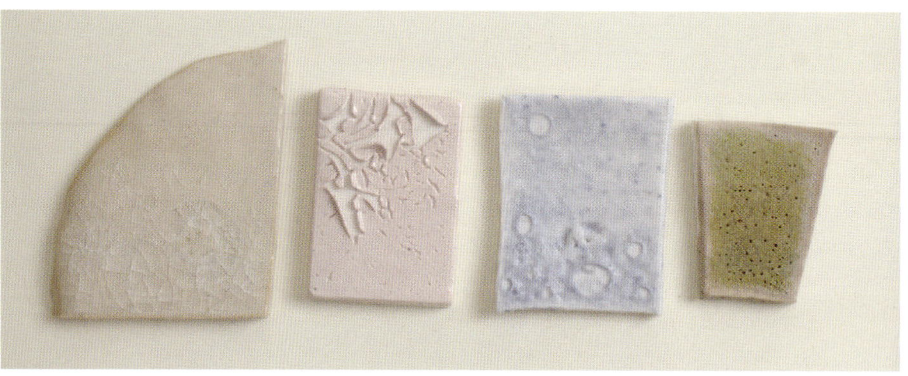

Crazing, crawling, blisters and pinholes in glazes.

Summary for adjusting glazes and special effects

To create special effects (or 'defects')

- To increase crazing, replace potash feldspar with soda feldspar or nepheline syenite or reduce silica in 5% increments.
- To cause crawling, replace dolomite with light magnesium carbonate and apply glaze more thickly.
- To encourage pinholes, dampen biscuit ware before applying glaze.
- To cause blisters, add silicon carbide 2% and titanium dioxide 5% to a matt glaze or an underlying slip.

To correct special effects (or 'defects')

- To correct crazing, add 5% silica or talc.
- To correct shivering, add 5% nepheline syenite, or reduce silica by 5%.
- To prevent crawling, apply glaze more thinly, reduce clay content, or sponge clean biscuit ware.
- To smooth pinholes, soak (hold at peak temperature) for 15–30 minutes at the end of firing.
- To avoid blisters, fire to a lower temperature, or apply glaze more thinly.

OPPOSITE TOP LEFT: Hollis Engley, *Small Crawled Cup*, thrown stoneware, STARworks New Catawba clay body, kiawe ash liner glaze, shino slip, wood-fired in Rose Esson-Dawson's kiln.

OPPOSITE TOP RIGHT: Amy Cooper, *Urchin Light*, slipcast porcelain, pierced, with shrink-and-crawl glaze containing light magnesium carbonate.

RIGHT: Jan Lewin-Cadogan, thrown stoneware with layered volcanic and barium glazes.

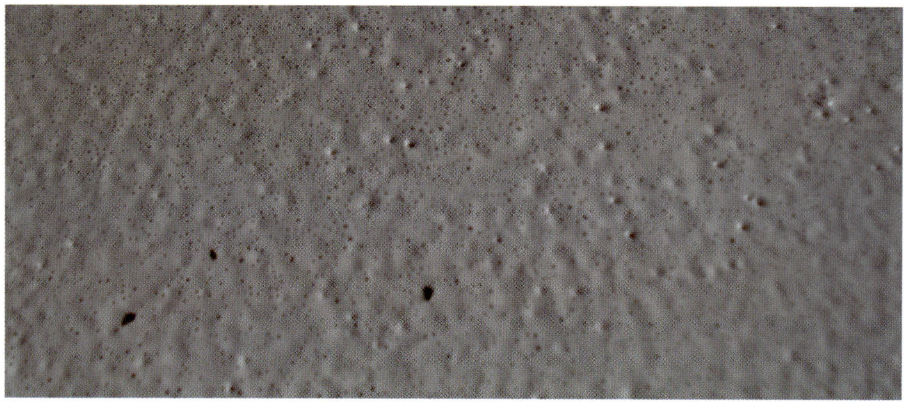

LEFT: Pinholed glaze. *Photo: Fay De Winter.*

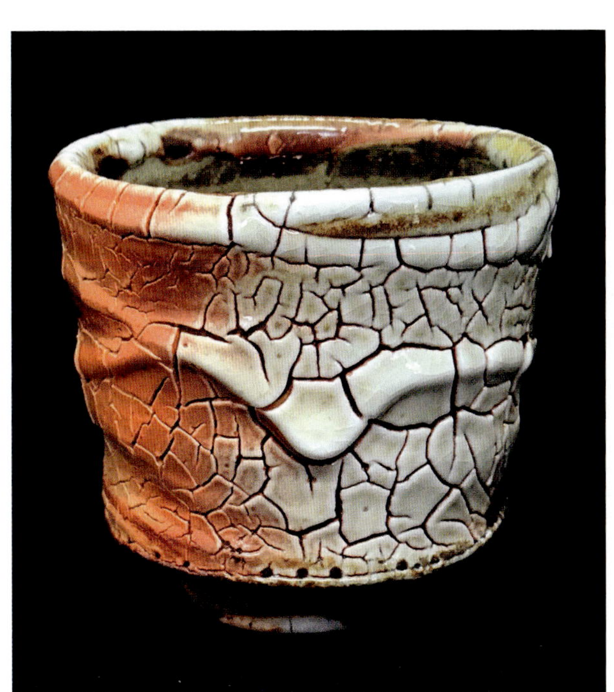

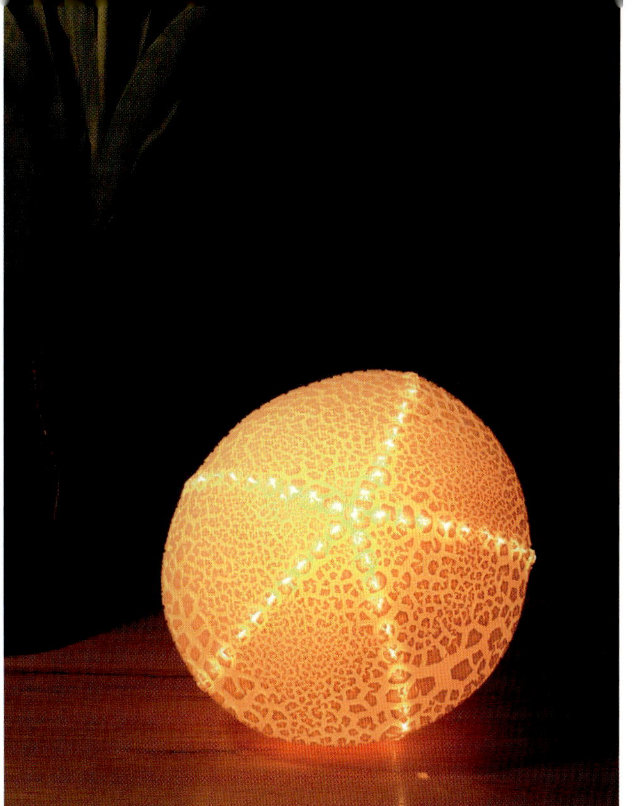

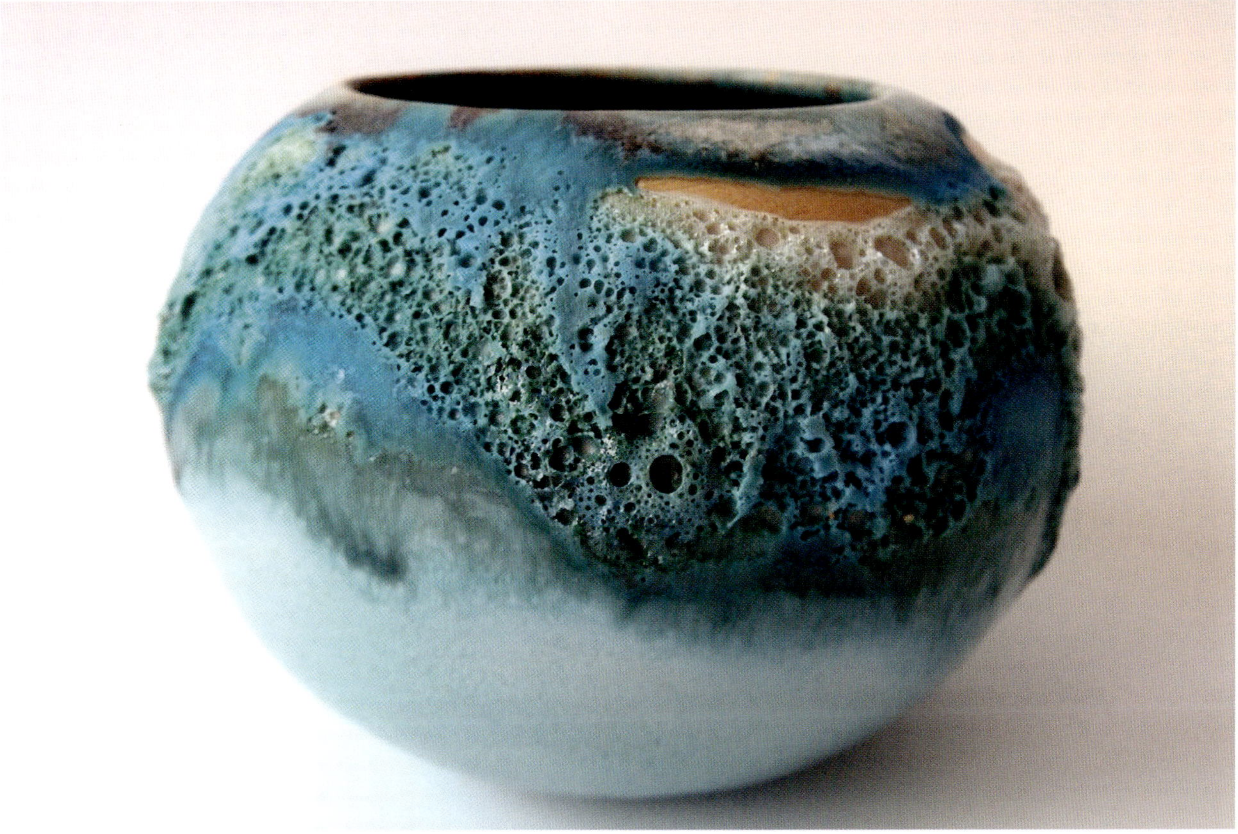

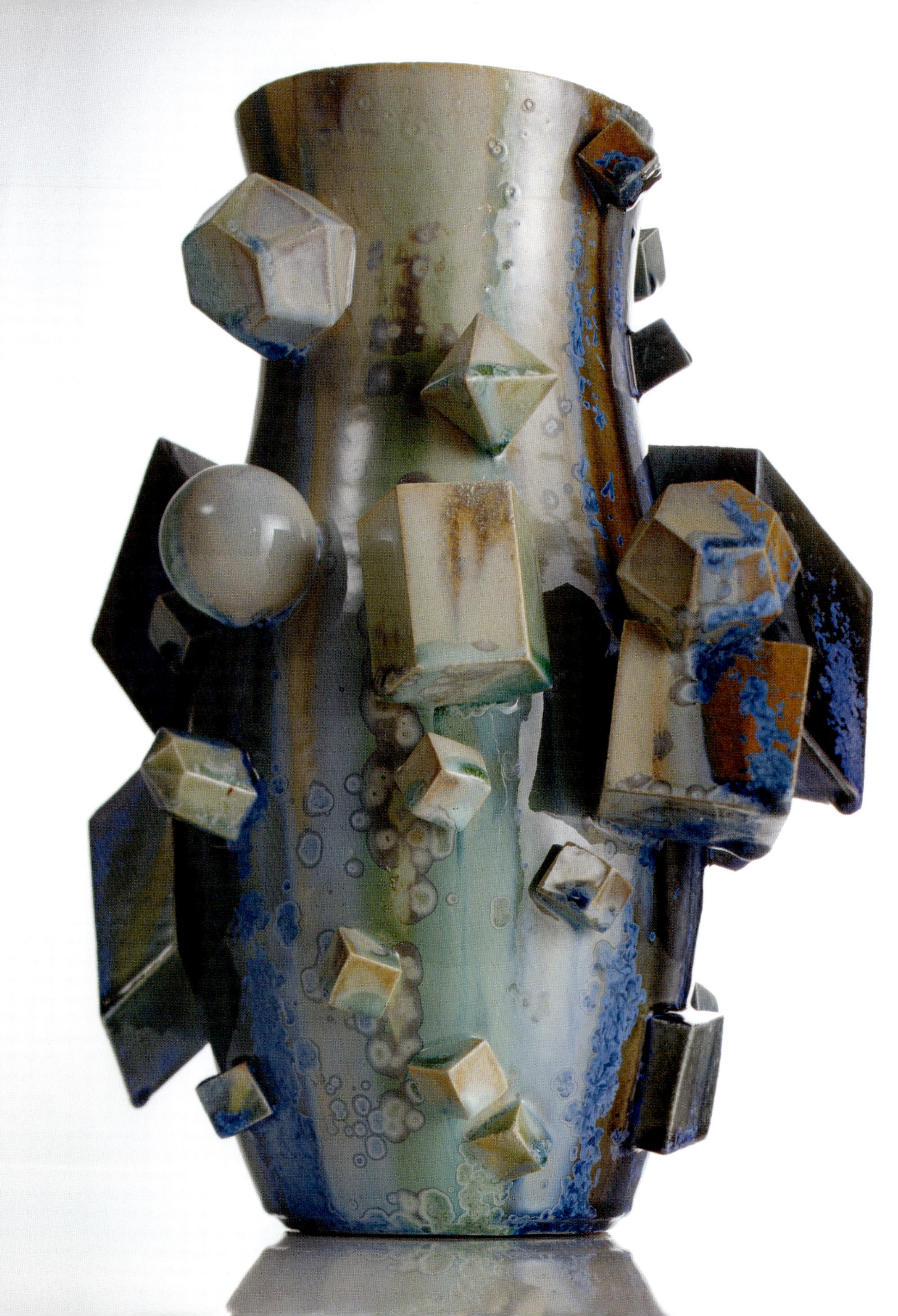

SECTION 2
Special-effect glazes

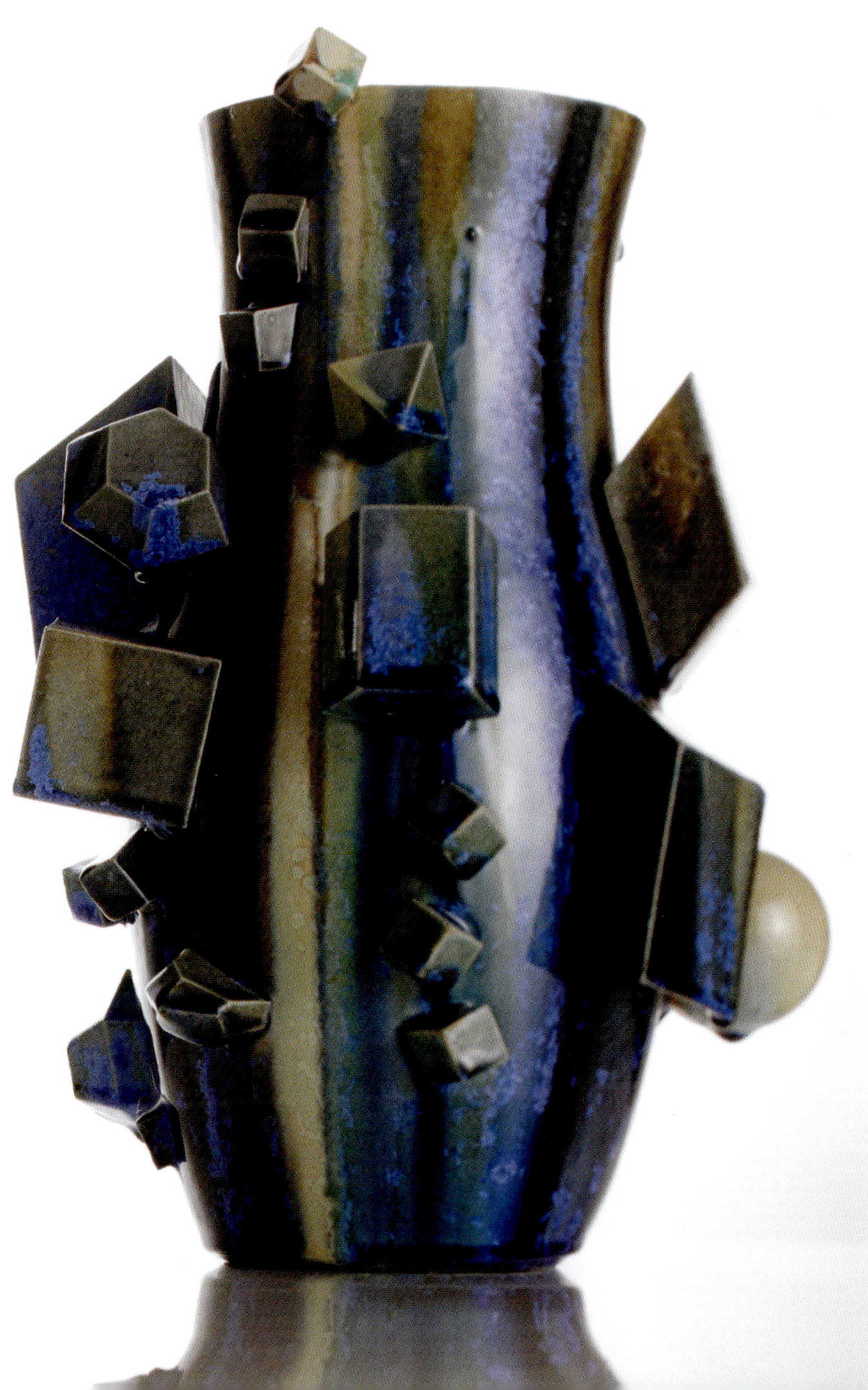

Kate Malone, *A Pair of Striped Magma Vases*, crystalline-glazed stoneware, 2018. The glazes are coloured in stripes with, from left, copper, nickel and cobalt oxides. Image courtesy of Adrian Sassoon, London. *Photo: Sylvain Deleu.*

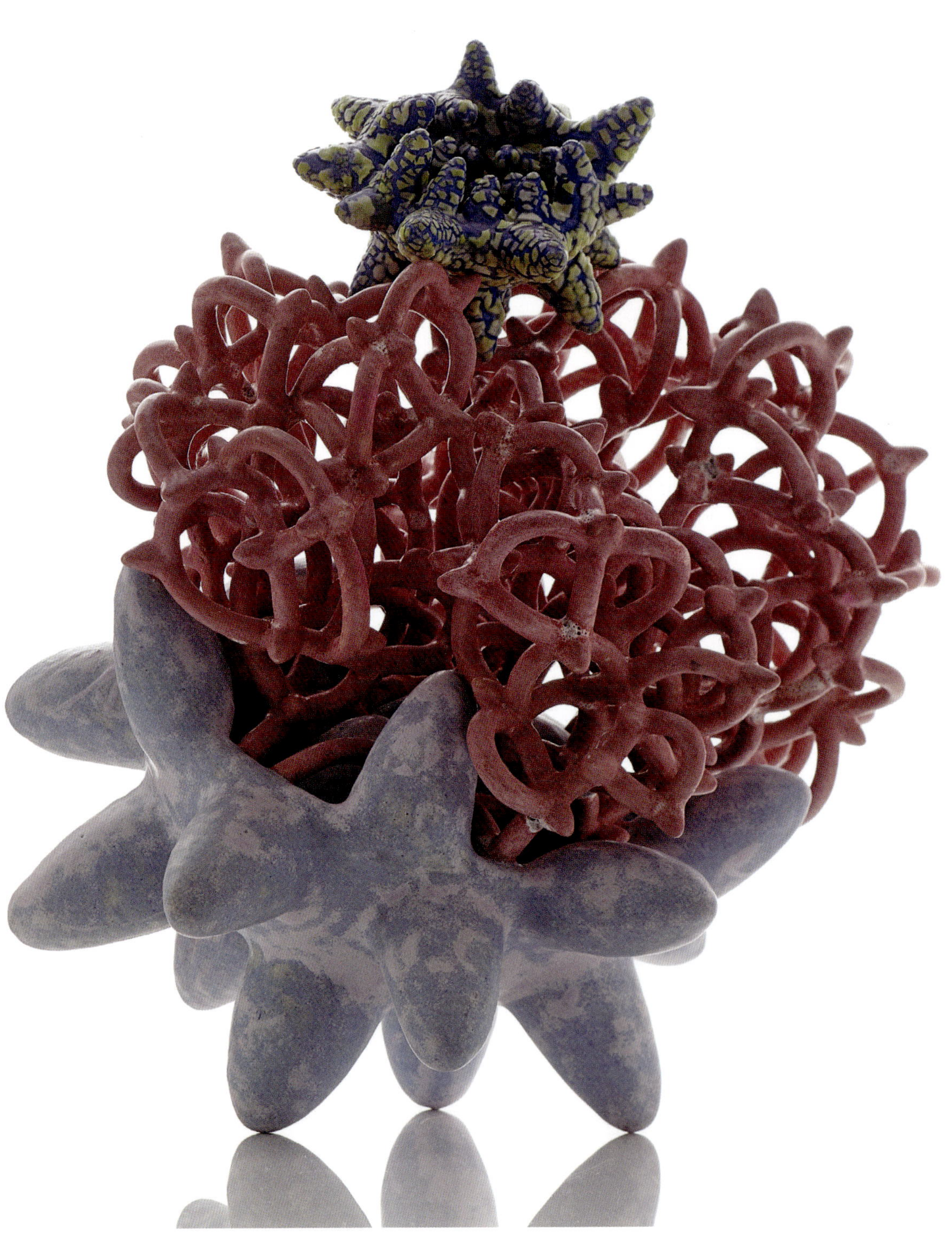

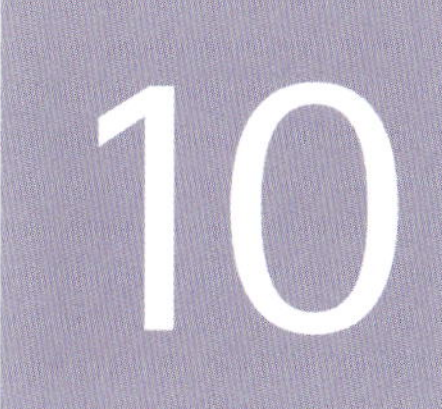

10 Special effects: the chemistry

Remember that glazes consist of three main components: the glass former silica, fluxes to melt the silica, and alumina sourced from clay to stiffen the melt so that it is not too runny. In the section on glaze stability and durability, we covered the choice of fluxes and how different flux combinations affect the glaze stability. Here we will look at the ratio of silica to alumina and see how different proportions affect the properties of the glaze, to determine whether it is glossy, matt or crystalline.

Stull map of special effects

To investigate special-effect glazes, we first need to understand how well-behaved, glossy glazes work. Plotted on a graph of silica against alumina (see glaze diagram 1 below), the glazes with the highest gloss lie on the dashed line representing a ratio of 1:8 alumina to silica (from the unity molecular formula, divide the number of molecules of silica by the number of molecules of alumina to get 8). This ratio of 1:8 represents the eutectic mixture of silica and alumina, the particular combination that has the lowest melting temperature.

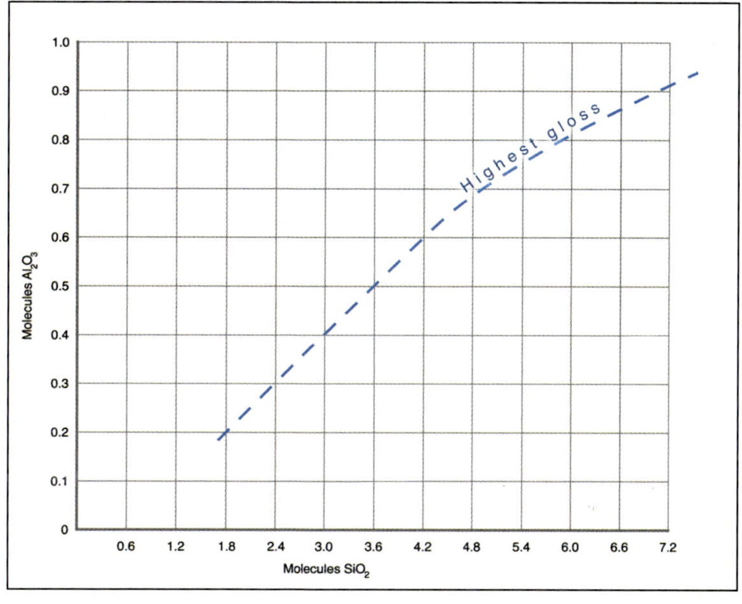

Glaze diagram 1

LEFT: Tessa Eastman, *Pollinating Creatures I & II*, 2016, 30 x 30 x 31cm (11¾ x 11¾ x 12¼ in.). Private Collection. *Sylvain Deleu Photography.*

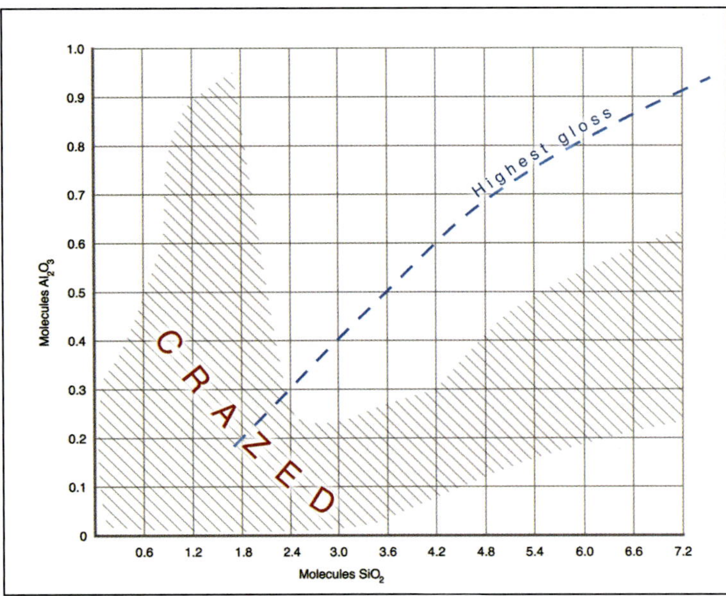

Glaze diagram 2

The graph was originally derived by R.T. Stull in 1912 using glaze tests with the fluxes potassia and calcia in the molecular ratio 0.3:0.7. For those combinations of fluxes, silica and alumina on a porcelain clay body, the glazes that crazed are shown using hatched lines (glaze diagram 2).

Crazing occurs when the expansion of the glaze is greater than that of the clay body. In other words, glazes in the hatched areas on the graph contained too little silica and/or alumina to counteract crazing caused by the high-expansion flux potassium oxide.

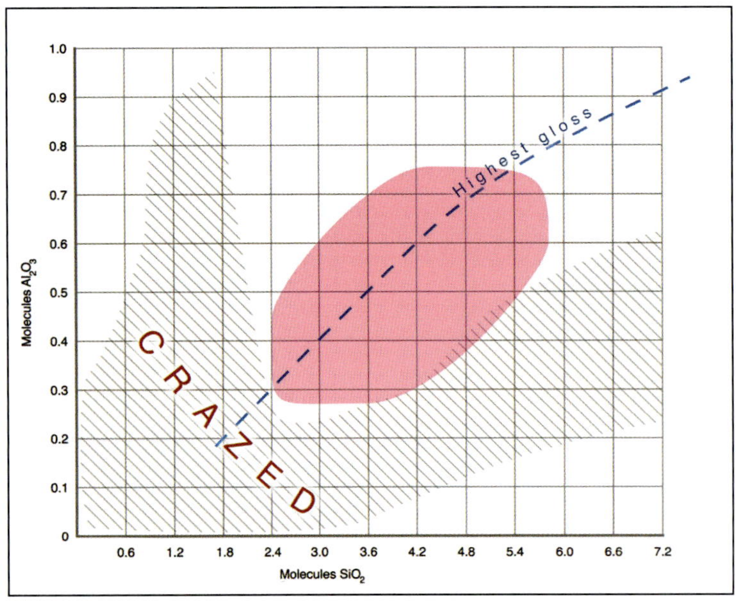

Glaze diagram 3

In the next diagram (glaze diagram 3), the pink area shows all the glazes within the limits for stable glazes (given by Cooper and Royle, 1984) for firings between cone 5 and cone 8 (these are suitable firing temperatures in electric kilns). These stable glazes are generally glossy and craze-free.

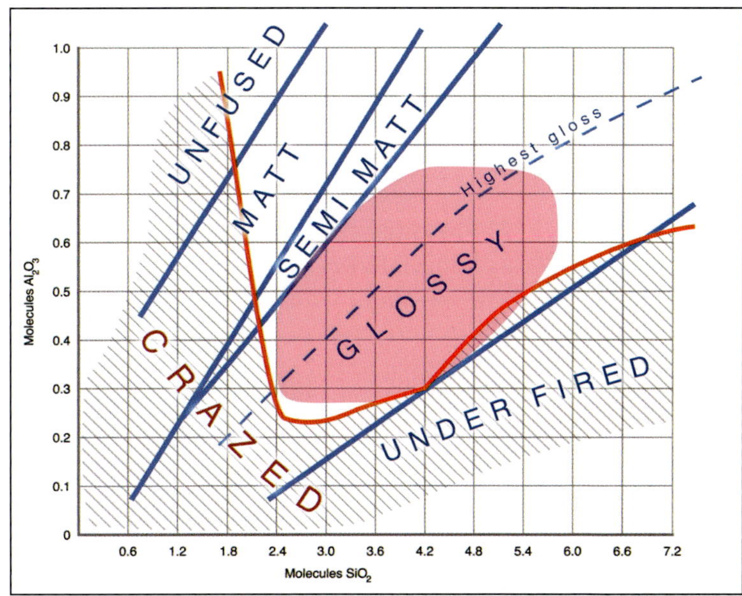

Glaze diagram 4

In this next diagram (glaze diagram 4), solid blue lines have been added to show matt glazes (1:4 alumina to silica) and semi-matt glazes (1:5 alumina to silica), as well as unfused and under-fired glazes which were not properly melted, owing either to high alumina or high silica. The data on the graph comes from glaze tests on porcelain fired to cone 11 by R.T. Stull. However, the same graph can be used for other clay bodies and firing temperatures, provided enough boron is used to melt the glaze at the chosen temperature (0.15 molecules boron oxide at cone 6).

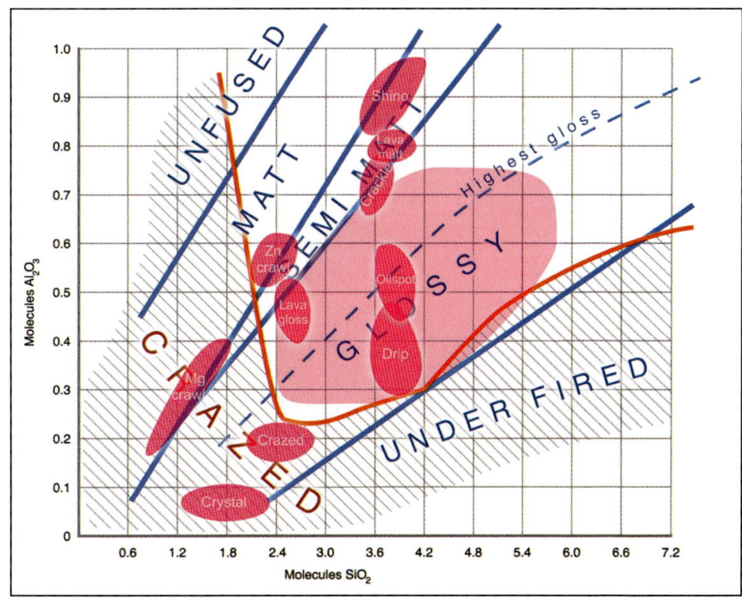

Glaze diagram 5

Finally, we come to the areas on the graph where different types of special effect glazes are found (glaze diagram 5). Many special-effect glazes are those that lie outside the boundaries or limits for stable, glossy, craze-free glazes. In this book we will explore each type of special effect and see where it lies on the map.

Stull map showing the alumina and silica range of various special-effect glazes. The effect of changing the alumina and silica in porcelain glazes fired to cone 11 with constant flux 0.3 K_2O and 0.7 CaO. The ratio of 1:5 alumina to silica gives a semi-matt glaze, while 1:8 gives a shiny glaze. The straight lines on the chart represent alumina:silica ratios of 1:4 (matt), 1:5 (semi-matt) and 1:12 (shiny, crazed glaze). The dashed line is 1:8 Al_2O_3:SiO_2 (bright, shiny glaze). The hatched area shows crazed glazes. The pale pink area shows Cooper and Royle's limits for stable glazes from cones 5 to 8. The special effects shown may move to the left for lower firing temperatures or to the right for higher firing temperatures. Data from R.T. Stull, 1912. Graphics by Henry Bloomfield.

Crazed and crystal glazes are low in both silica and alumina, while crawl and shino glazes often have high alumina. Volcanic or lava glazes are usually viscous, semi-matt glazes with moderately high alumina, but the addition of silicon carbide is also an important factor in causing eruptions of gas. In contrast, runny, drippy glazes are relatively low in alumina and have low viscosity in the melt. Oil-spot glazes are more fluid in the melt than volcanic glazes but still rely on outgassing for their spotted effect. The important ingredient for most oil-spot glazes is the addition of around 6–10% iron oxide, which releases oxygen at stoneware firing temperatures. Vanadium pentoxide, manganese dioxide and copper oxide also release oxygen at high temperatures and can be used to make pitted and spotted glazes.

This diagram does not explain everything about special-effect glazes (such as the action of silicon carbide) but it is a useful tool to start thinking about the relationships between them and how variations in glaze ingredients can lead to variations in glaze effects.

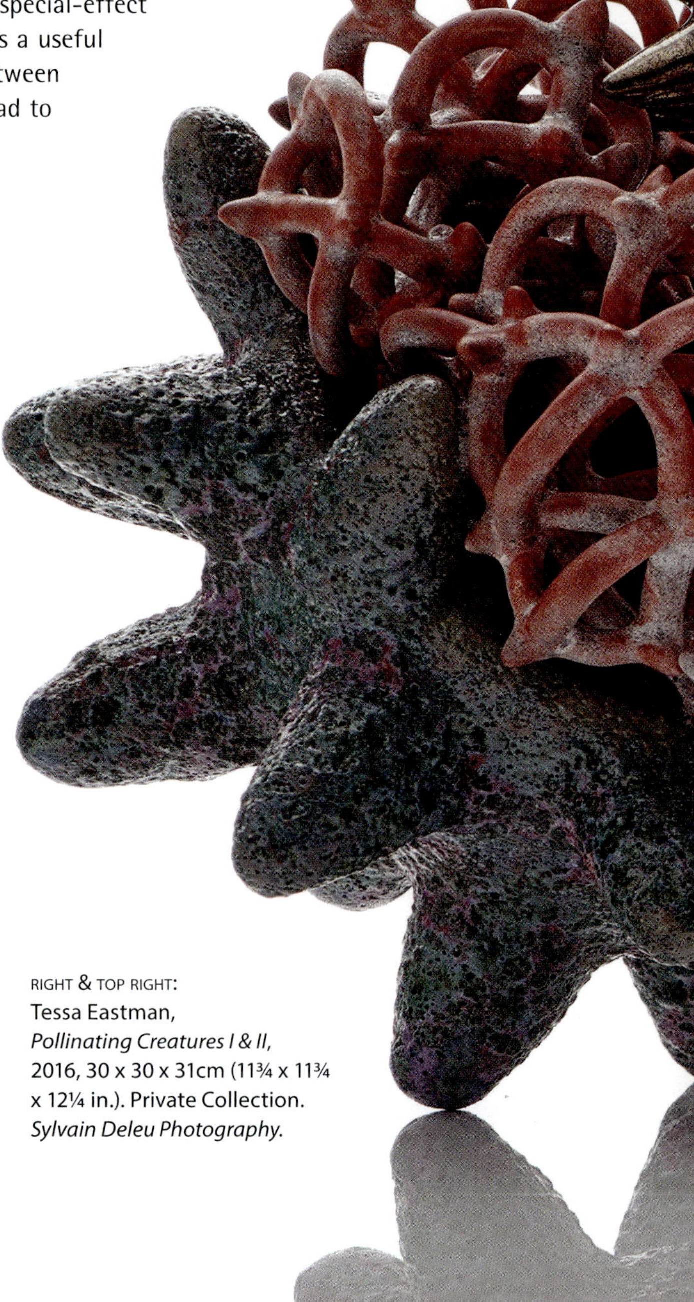

RIGHT & TOP RIGHT:
Tessa Eastman,
Pollinating Creatures I & II,
2016, 30 x 30 x 31cm (11¾ x 11¾ x 12¼ in.). Private Collection.
Sylvain Deleu Photography.

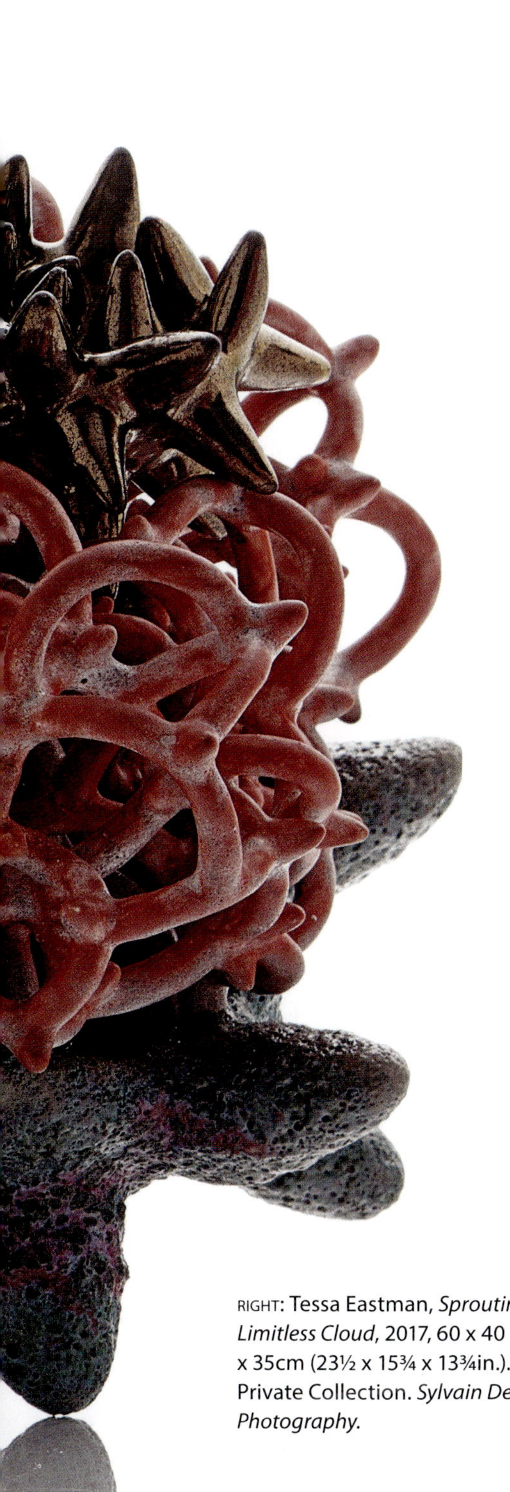

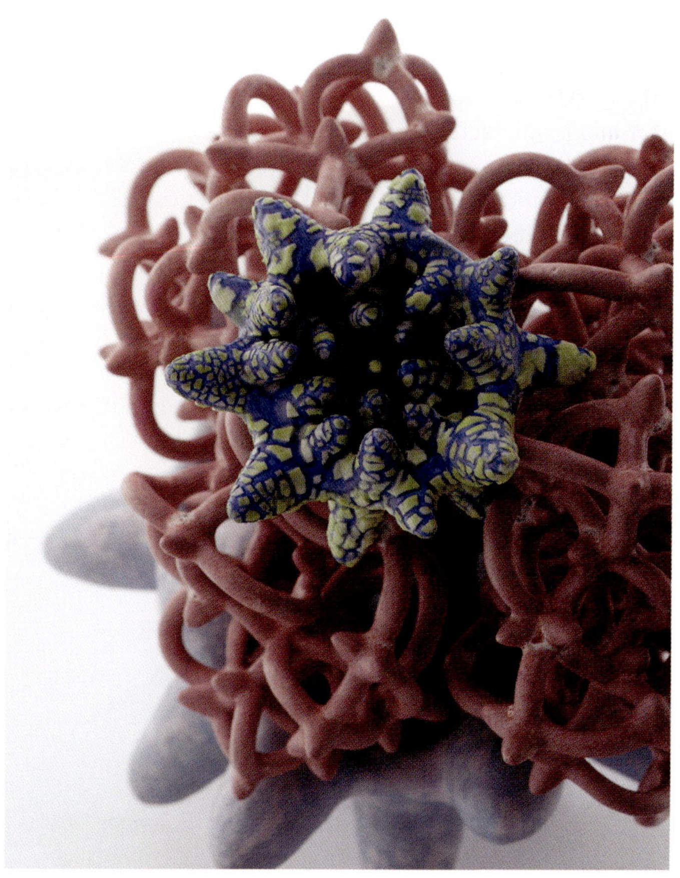

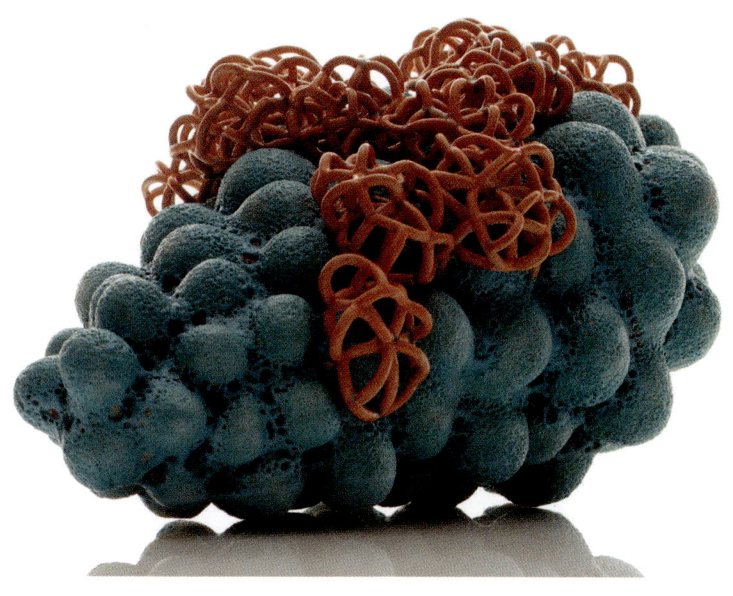

RIGHT: Tessa Eastman, *Sprouting Limitless Cloud*, 2017, 60 x 40 x 35cm (23½ x 15¾ x 13¾in.). Private Collection. *Sylvain Deleu Photography.*

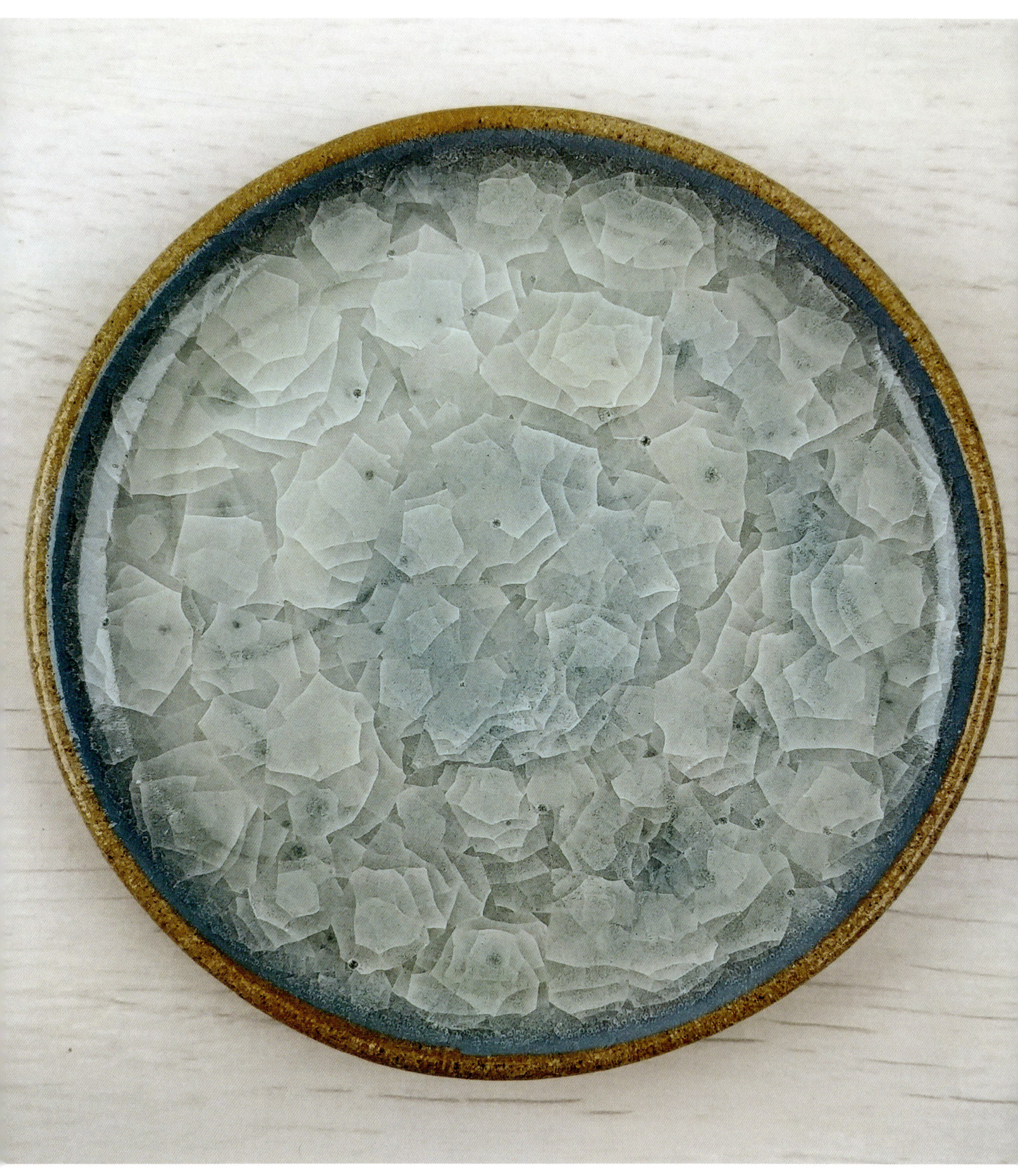

11

Crackle glazes

Crazing or crackle is technically a defect where the glaze is too small for the clay body. On cooling in the kiln, the glaze contracts more than the clay and a network of fine cracks develops. Cracks can continue to appear over weeks and months, particularly if there are sudden temperature changes such as those which occur when the pot is put in the oven or dishwasher. Crackle is often used as a decorative effect in ceramics from the Far East, but the cracks can greatly weaken the pot, causing it to crack and break more readily than a pot with an uncrazed glaze. Crazing on earthenware pots can cause them to leak, as the fired clay body remains porous and water can seep through. The cracks can also harbour dirt and bacteria, so are not ideal on functional pots, particularly in earthenware. Crackle glazes are often high in sodium and may not be chemically durable (i.e. not resistant to attack by acids in food or alkaline soap in the dishwasher). Some potters use a different glaze on the inside to avoid this problem, but this must be done carefully – a glaze with a different expansion can cause the work to crack.

Crazing occurs when the glaze has a higher thermal expansion than the clay body. If the glaze and clay body have very different rates of expansion, the glaze fit will be poor and crazing may occur. The extent of crazing will depend on the stress or mismatch between the clay and glaze. The brittle nature of glaze means it is not able to stretch, so instead it forms a network of cracks to release the stress. This can occur over time and is responsible for the pinging sound sometimes heard when pots are removed from the kiln. Crazed or crackle glazes can be achieved by adding glaze materials which have a relatively high expansion. These materials include soda feldspar, nepheline syenite and high alkaline frit. Guan (kuan) and Snowflake crackle are types of Chinese crackle glaze, applied thickly, that give a deep crackle pattern.

How to adjust crackle without changing the appearance of the glaze

We will learn about crackle by first considering the options available for eliminating crazing. While gaining an understanding of how to either reduce or increase crazing, at the same time we will learn how to make crackle glazes.

Joe Thompson, *Old Forge Creations*, thrown stoneware dish, Snowflake crackle glaze over blue slip, fired to cone 6, dia: 12cm (4¾in.). Recipe given on p.83.

Adding silica and clay

There are several ways to correct crazing. However, changing only one material may change the appearance of the glaze, making it more glossy or matt. A reliable method is to increase both the silica (flint or quartz) and clay in the ratio 1.25:1 silica to clay. This ratio comes from a series of porcelain glaze tests made by R.T. Stull in the USA in 1912. He measured the effect on the glaze of changing the alumina and silica content. He found that a molecular ratio of 1:5 alumina to silica gives a matt glaze, while 1:8 gives a shiny glaze. Alumina is found in clay, which is added to glaze to stiffen it in the melt and prevent it running off the pot. Clay contains silica as well as alumina, and the ratio 1:8 alumina to silica molecules converts to a ratio by weight of 1:1.25 clay to silica. To change your glaze recipe, add increments of 1% clay and 1.25% silica until the crazing disappears. This has the effect of moving along the dashed blue line in the graph and crossing the red line from the crazed to the glossy, uncrazed area. You will need to multiply the percentage by the total weight; for example, in a 100g dry batch of glaze, add 1g china clay and 1.25g quartz, then wet-sieve and test on a pot to see if the crazing is reduced. If not, add increments of 1g clay and 1.25g quartz, up to 4g clay and 5g silica and test again. The cracks should move further apart the more clay and silica are added. Test tiles are often too small an area to see crazing, so you may need to test on a larger bowl or plate. Crazing is also affected by the glaze thickness. A thinner application of glaze or the addition of more water is sometimes enough to reduce the crazing, but obviously for a special effect we may want it to craze, so to increase crazing, reduce the silica and clay in the glaze and apply the glaze more thickly.

RIGHT: Linda Bloomfield cup and saucer, runny craze-free turquoise glaze fired to cone 8.

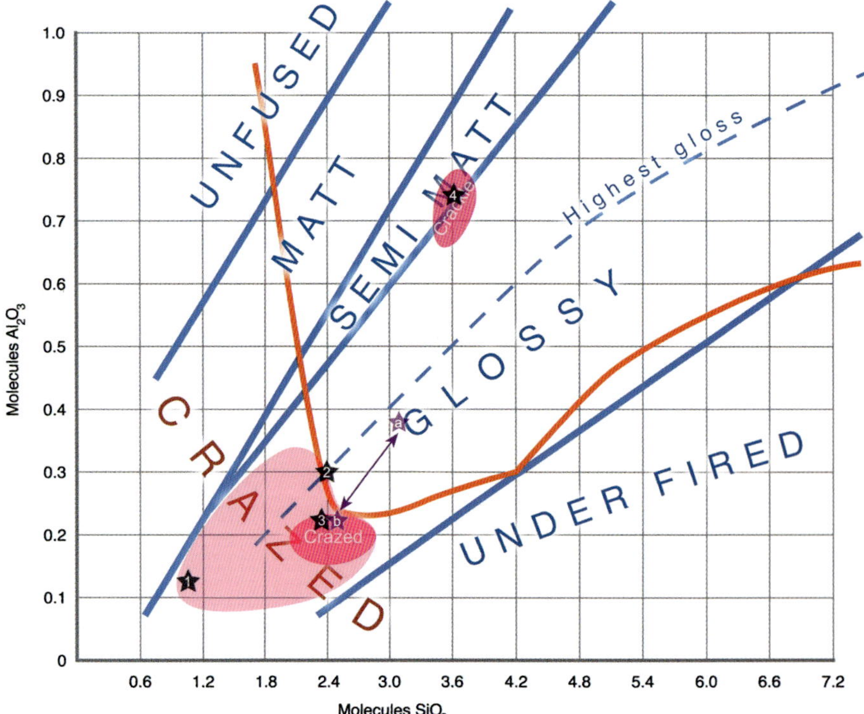

Stull map of alumina and silica in porcelain glazes fired to cone 11 with constant flux 0.3 K_2O and 0.7 CaO. By increasing the alumina and silica in the molecular ratio 1:8, you can move a glaze from the crazed area (b) to the glossy, uncrazed area (a) and vice versa. Glossy, crazed glazes are found in the bottom left-hand corner, while Snowflake crackle is on the semi-matt 1:5 line (see Kuan glaze opposite). The numbers refer to glazes on pp 83–7. The ratio of 1:5 alumina to silica gives a semi-matt glaze, while 1:8 gives a shiny glaze. The straight lines on the chart represent alumina:silica ratios of 1:4 (matt), 1:5 (semi-matt) and 1:12 (shiny, crazed glaze). The dashed line is 1:8 Al_2O_3:SiO_2 (bright, shiny glaze). Data from R.T. Stull, 1912. Graphic by Henry Bloomfield. Thanks to Matt Katz of Ceramic Materials Workshop.

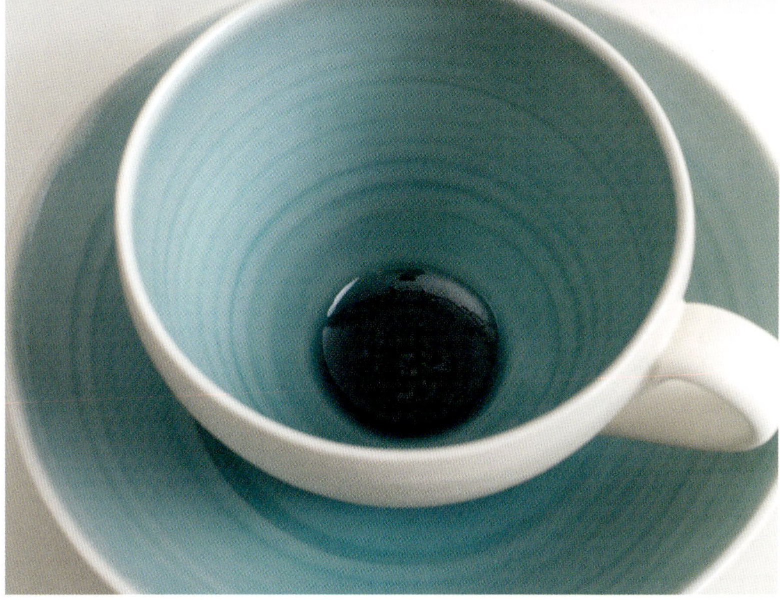

a Runny craze-free turquoise glaze, cone 6–8, 1240–1260°C (2264–2300°F)

This glaze (a) is in the glossy area on the graph (Na_2O 0.25, CaO 0.75, B_2O_3 0.38, Al_2O_3 0.38, SiO_2 3.1)

Soda feldspar 47
Calcium borate frit 16
Whiting 14
China clay 5
Quartz 18
+
Copper oxide 1

Recipe numbers correspond to numbers marked on Stull map opposite.

b High-alkaline turquoise crackle glaze, cone 6, 1240°C (2264°F).

This glaze (b) is in the crazed area on the graph (Na_2O 0.7, CaO 0.3, Al_2O_3 0.23, SiO_2 2.5).

(Not food-safe.)

Soda feldspar 15
High alkaline frit 47
Lithium carbonate 2
Whiting 6
Quartz 18
China clay 10
+
Copper oxide 2

4 Kuan (Snowflake) crackle glaze, cone 6–8, 1240–1260°C (2264–2300°F). Glaze 4 on the Stull map.

(Apply very thickly.)

Nepheline syenite 80
Whiting 5
Borax frit 15
+
Bentonite 2

Adding fluxes

Adding too much silica and clay may make the glaze appear underfired, dry and matt. Another way to correct crazing is to add a low-expansion flux material such as talc, which is magnesium silicate. Both magnesium oxide and silica have low expansion; both will decrease the expansion and contraction of the glaze during cooling, to help prevent crazing. Talc is a better choice than dolomite, which contains both magnesium and calcium oxide, the latter of which has a relatively high expansion (although lower than that of sodium or potassium). If you look at the fluxes on the next page, listed in order of decreasing expansion, with high expansion at the top and low expansion at the bottom, you can choose the most appropriate low-expansion flux to add to your glaze. It is simpler to increase a material already in the glaze than to add a new one. Stoneware glazes can have the addition of 5% talc or zinc oxide.

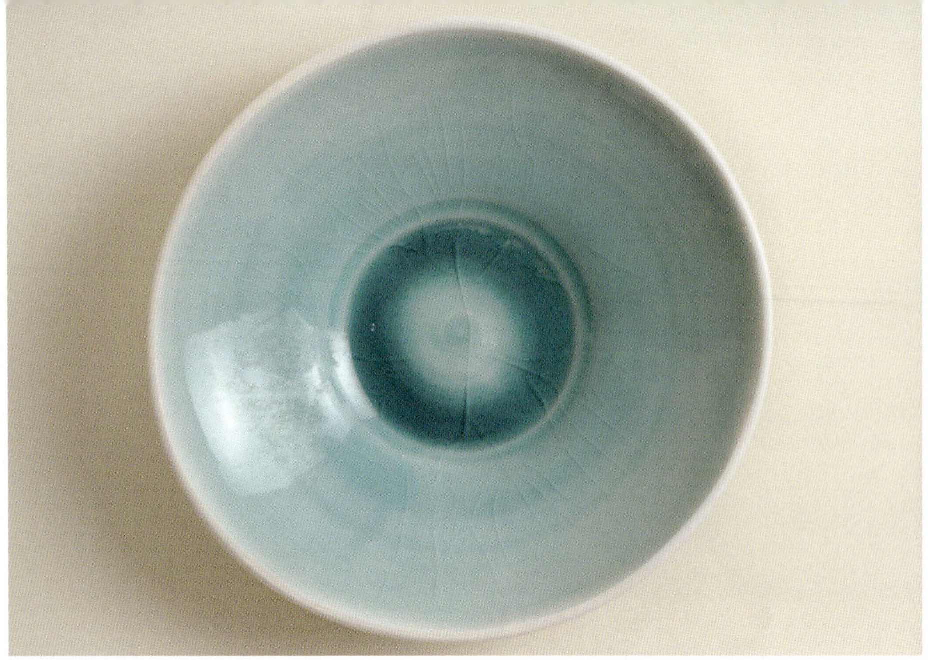

Linda Bloomfield, porcelain bowl with runny turquoise glaze, fired to cone 8. The crazing was caused by a change in the feldspar composition.

However, these fluxes may affect the colour of chrome green and chrome-tin pink glazes. Earthenware glazes can have the addition of a flux containing boron, which has a very low expansion and can be added in the form of calcium borate frit (or colemanite). Opacifiers and colouring oxides such as zirconium silicate and tin oxide also help to reduce crazing.

High expansion

↑

Sodium oxide (in nepheline syenite and high-alkaline frit)
Potassium oxide (in potash feldspar)
Calcium oxide, strontium oxide
Barium oxide
Titanium oxide, lead oxide
Lithium oxide (in lithium carbonate and spodumene)
Zinc oxide
Magnesium oxide (in talc)
Tin oxide, zirconium oxide
Alumina (in clay)
Silica
Boron oxide (in calcium borate frit)

Low expansion

However, adding a flux will often make the glaze more runny. Alternatively, you can remake the glaze, reducing the high-expansion materials such as feldspar, or substituting low-expansion materials for the high-expansion ones, such as a lithium feldspar for soda feldspar or calcium borate frit for high-alkaline frit. It is a good idea to change one material at a time to keep a clear idea of the effect of each individual material on the glaze. On the other hand, if you want to induce crazing, add a high-expansion flux such as nepheline syenite or a high-expansion frit such as high-alkaline frit (or Ferro frit 3110). Increasing the fluxes has the same effect as reducing the clay and silica in the glaze. Glazes consisting of 50% or more feldspar or nepheline syenite are likely to craze. Snowflake crackle (kuan) glazes are semi-matt glazes consisting mostly of nepheline syenite and are applied very thickly (see kuan glaze recipe on p.83). On the Stull map, these Snowflake crackle glazes lie on the 1:5 semi-matt line, while glossy crackle glazes low in silica and alumina lie in the crazed area in the bottom left-hand corner.

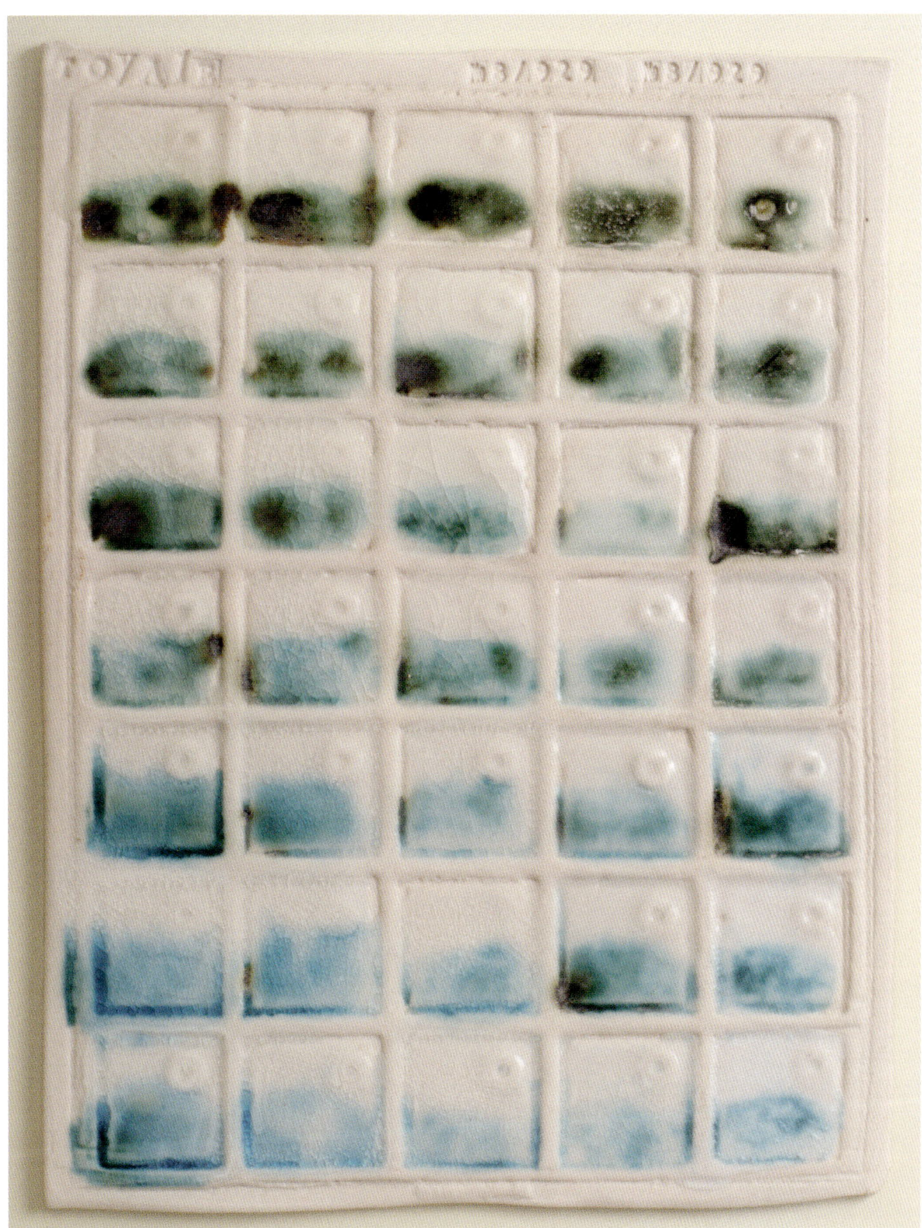

Biaxial grid with high-alkaline glaze and copper oxide brushed on porcelain. Silica increases from 0-50% left to right. Clay increases from 0–40% bottom to top. The high-alkaline fluxes (Na_2O+K_2O+Li_2O 0.7, CaO 0.3) remain constant throughout. The only uncrazed tile is in the centre row, fourth from left (14 kaolin, 34 silica). Interestingly, the copper changes from turquoise to green as more clay is added, which is why transparent turquoise glazes are often crazed. Grid made by May Luk and the London Potters glaze group.

Changing the clay body or firing temperature

If you do want crazing to happen, try lowering the firing temperature. Avoid adding extra silica or sand to the clay body. Conversely, ways to correct crazing include changing to a different clay body which better fits the glaze, adding silica to the existing clay body or increasing the firing temperature. Commercial clay bodies often already have silica added to prevent crazing. In stoneware, the addition of silica sand to the clay body can help prevent crazing. In earthenware, bisque firing to a higher temperature can eliminate crazing.

Summary

To increase crazing

Reduce silica and clay in the ratio 1:1.25 clay to silica by 5% silica and 4% clay
Add nepheline syenite or lithium carbonate
Substitute high-alkaline frit for borate frit
Apply glaze very thickly
Reduce firing temperature

To reduce crazing

Increase silica and clay by 5% silica and 4% clay
Add 5% talc or zinc oxide
Substitute lithium feldspar for sodium feldspar
Substitute borate frit for high-alkaline frit
Apply glaze thinly
Increase firing temperature

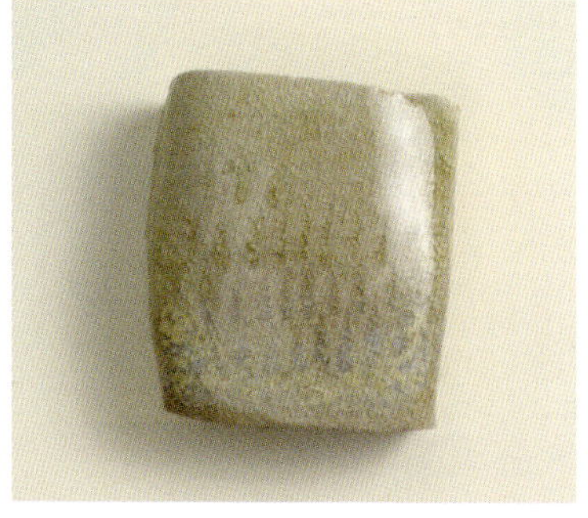

Simple runny ash glaze, cone 9, 1280°C (2336°F), on stoneware.

Potash feldspar 50
Wood ash 50
+
Bentonite 2

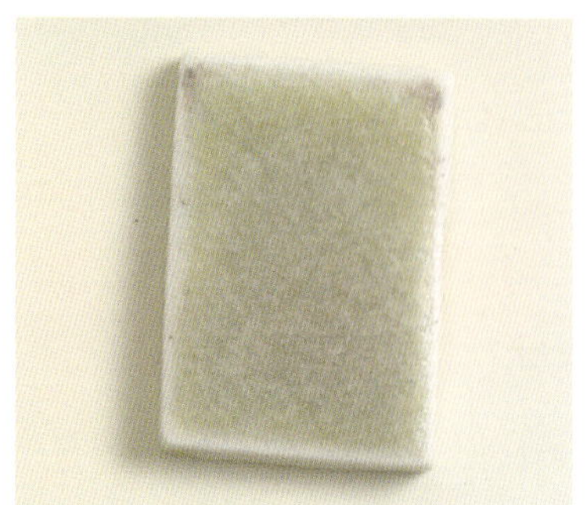

Semi-matt crackle ash glaze, cone 8, 1250°C (2282°F), on porcelain.

(The aluminium powder causes local reduction even when fired in an electric kiln.)

FFF feldspar 50
Wood ash 50
+
Aluminium powder 0.5

Recipe numbers correspond to numbers marked on Stull map on p.82.

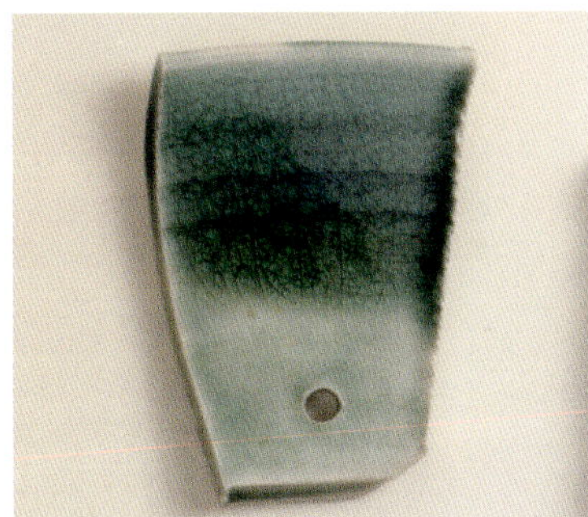

1 Runny green ash crackle glaze, cone 9, 1280°C (2336°F) on porcelain. Glaze 1 on the Stull map.

Wood ash 60
Feldspar 40
+
Copper oxide 1

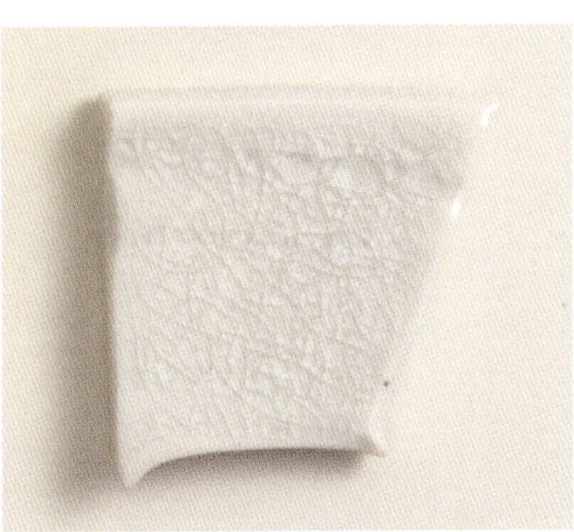

Transparent crackle glaze, cone 5–8, 1200–1260°C (2192–2300°F)

(Apply thickly. Not food-safe.)

Nephelene syenite 26
Whiting 12
Lithium carbonate 10
Barium carbonate 3
Calcium borate frit 5
Flint 44

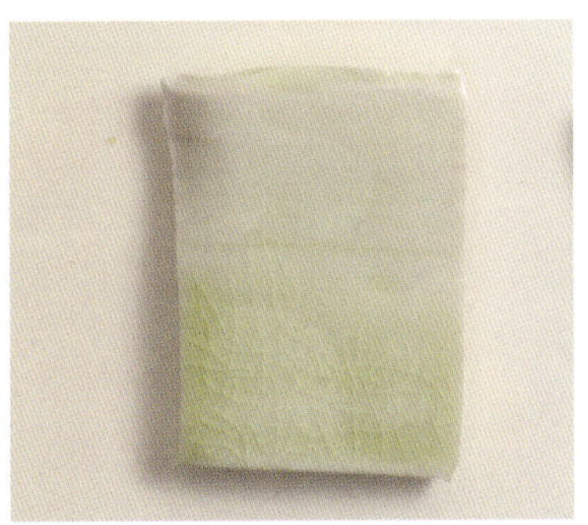

2 Chartreuse crackle glaze cone 6–8, 1240–1260°C (2264–2300°F). Glaze 2 on the Stull map.

(Apply thickly. Not food-safe.)

Soda feldspar 45
Calcium borate frit 15
Whiting 14
China clay 5
Quartz 17
Lithium carbonate 5
+
Chromium oxide 0.2

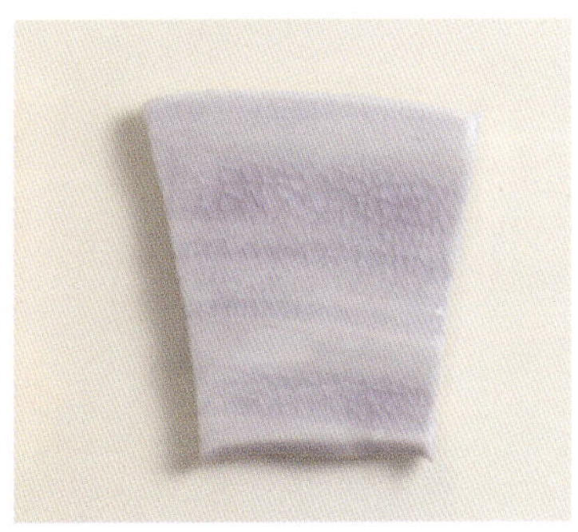

3 Violet crackle glaze, cone 6–9, 1240–1280°C (2264–2336°F). Glaze 3 on the Stull map.

(Apply thickly. Not food-safe.)

Nepheline syenite 24
Strontium carbonate 18
Lithium carbonate 10
Whiting 3
China clay 6
Flint 31
+
Neodymium oxide 8

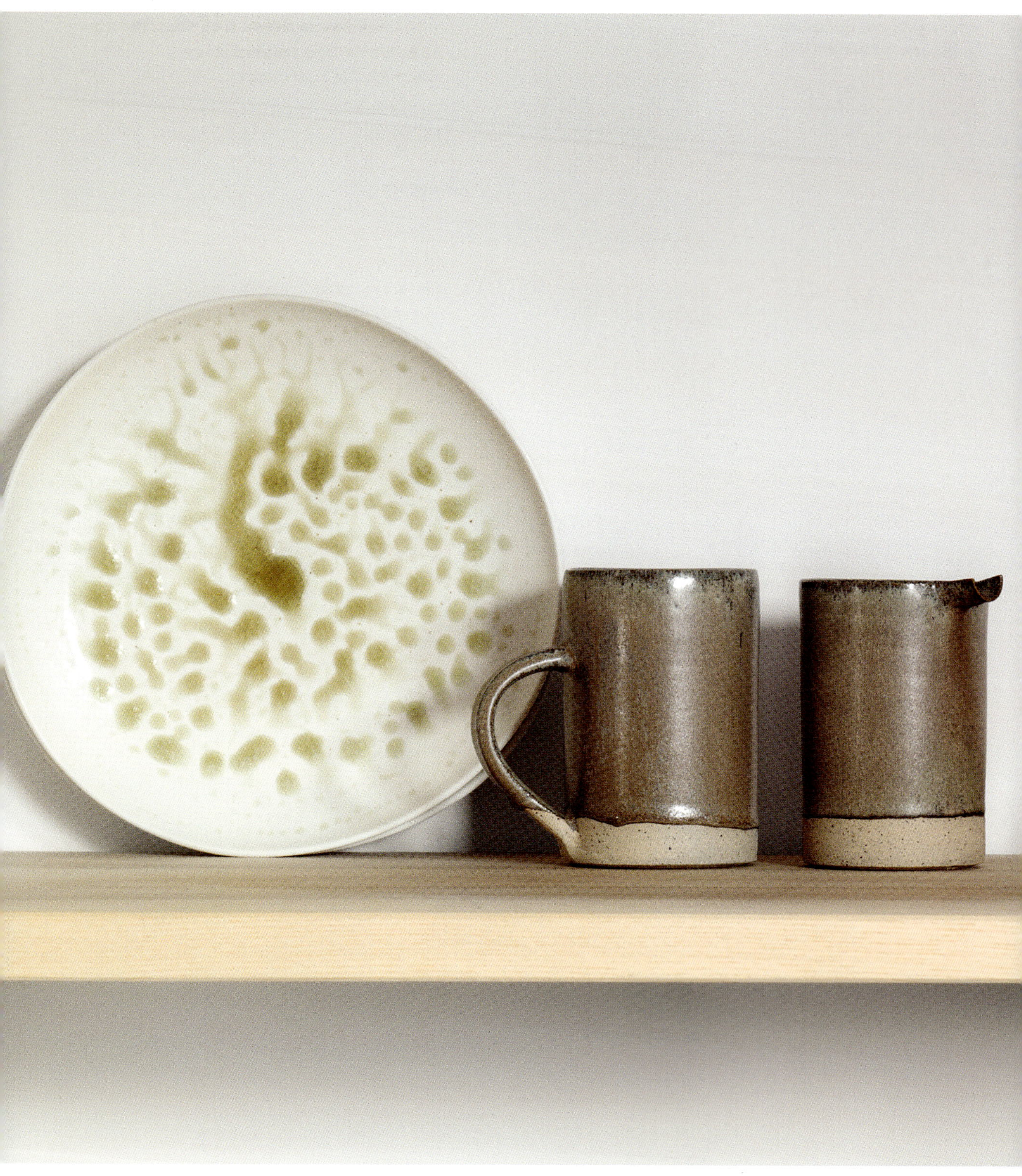

12

Ash glazes

Wood ash can be made into a runny glaze on its own or with the addition of some clay or feldspar and is one of the most widely available materials free to potters, besides locally dug clay. Wood ash is mostly calcium, silica and potassium; a full analysis is given later in this section. Ash can be collected from your hearth and used as it is, although I prefer to process it before using. First soak the ash in water in a plastic bucket or dustbin for several days. Remove the water (wear rubber gloves as it is highly caustic, containing dissolved potassium hydroxide, and will be yellow, with a slippery feel) and replace with fresh water. Repeat several times until the water remains clear. You can then sieve the wet ash through a coarse kitchen sieve to remove pieces of charcoal, and any nails or grit. It can be sieved again through a 60 or 80 mesh sieve and then dried in a biscuit-ware bowl before use in a glaze recipe. When mixing and using wood-ash glazes, wear rubber gloves, as the water will be alkaline, even after the ash has been washed. If not washed, it will still contain soluble alkali and will be very caustic. Ash glazes should be applied more thickly than other glazes as any remaining carbon in the ash burns away during firing. Apply more thinly on the outside of pots, as the glaze may run, and always test first.

Wood ash from a single type of wood will give subtle transparent colours, particularly if fired in reduction, where a range of greens can be obtained, from apple to olive green (see bowls by Akiko Hirai and Richard Batterham). Pine ash will give darker colours as it contains more iron and manganese. In oxidation, the fired colours vary from tan yellow to brown (see *Wilder* plate by Pottery West, shown opposite). Ash glazes can also be coloured using cobalt, chromium or copper oxides (see bowls by the author overleaf), but many potters prefer to bring out the natural colours of the ash.

LEFT: Ash-glazed porcelain plate, willow ash, fired to 1280°C (2336°F) in oxidation, dia: 21.5cm (8¼ in.). From *Wilder* range by Pottery West. *Photo: Jules Lister for Labrador 2017.*

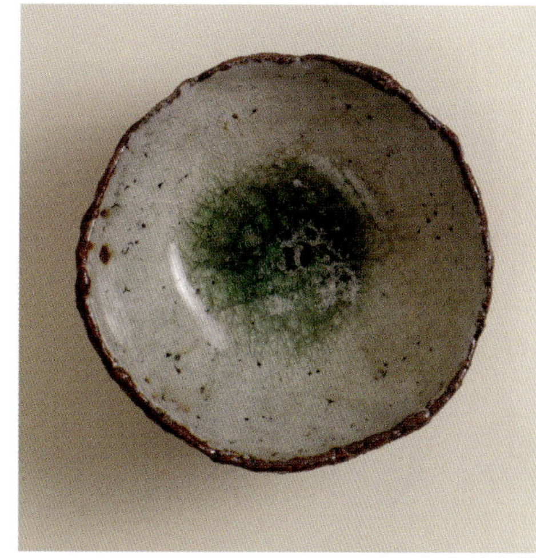

RIGHT, TOP: Akiko Hirai, Pond bowl, handbuilt stoneware with white slip and wood-ash crackle glaze, fired in reduction.

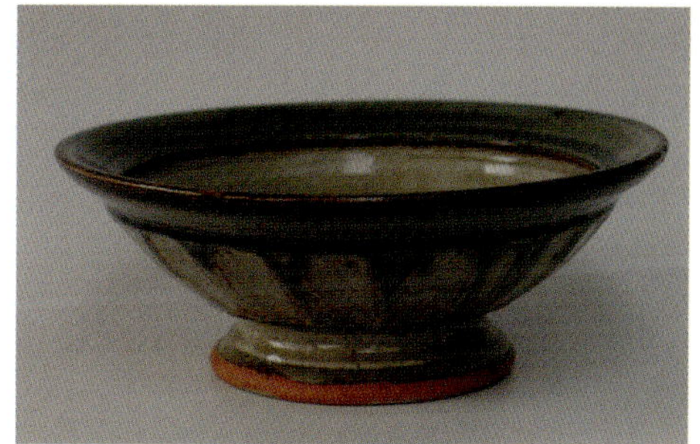

RIGHT, BOTTOM: Richard Batterham ash-glazed footed stoneware bowl, fired in reduction. Photo by Henry Bloomfield.

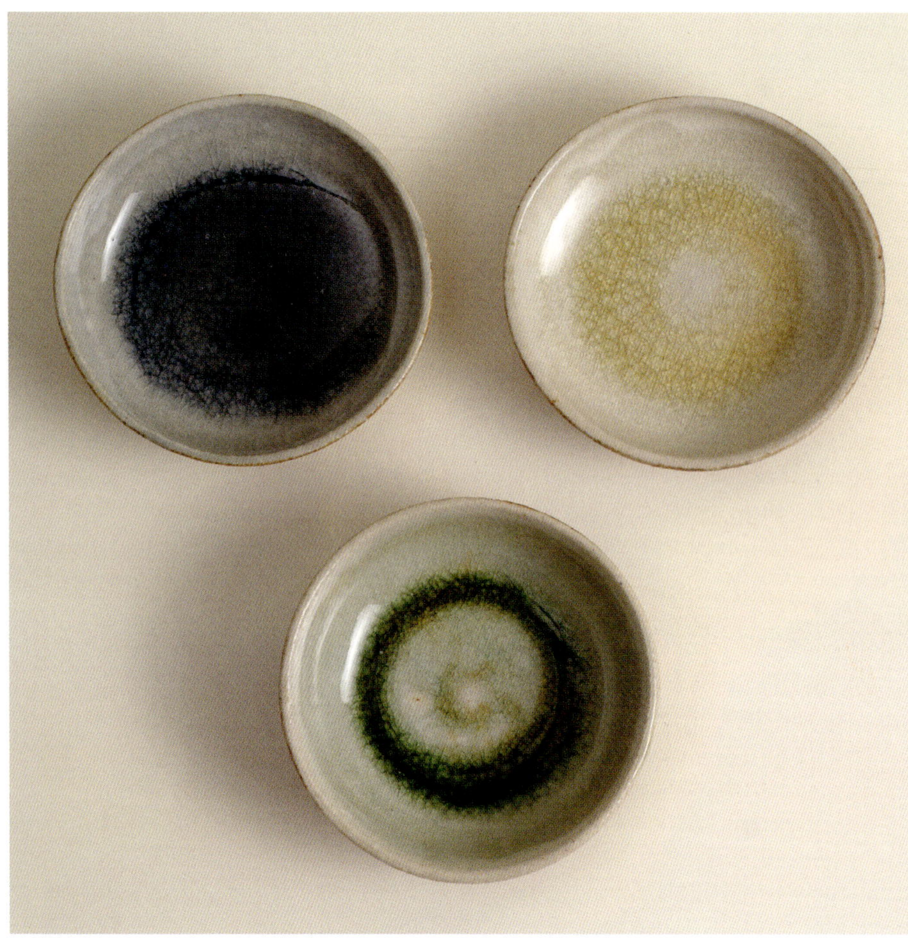

Linda Bloomfield, bowls. Simple ash glaze – see recipe p.86. Plain ash glaze (top right), coloured with cobalt (top left) and copper oxide (bottom), stoneware with white slip, fired in oxidation to 1280°C (2336°F).

BELOW: Robert Hunter, Mallorcan thrown grogged stoneware bottle with local olive wood ash glaze, reduction-fired.

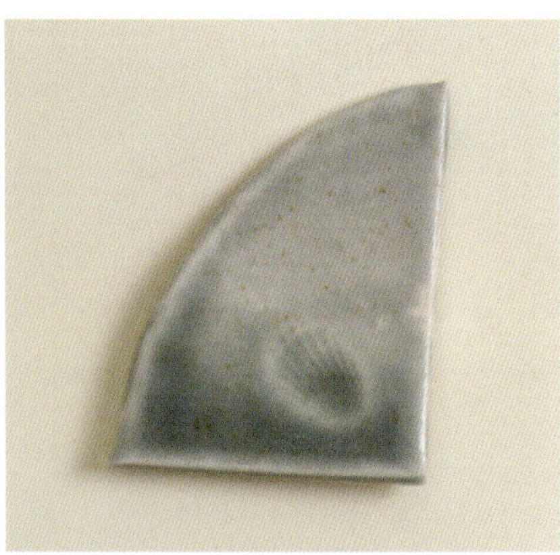

Nuka glaze (Stephen Parry), cone 10, 1300°C (2372°F), reduction.

Straw ash 80
Ball clay 20

Transparent green ash glaze on porcelain, cone 9, 1280°C (2336°F).

Wood ash 30
Potash feldspar 30
China clay 17
Whiting 4
Flint 21
+
Copper oxide 0.5
Cobalt carbonate 0.1

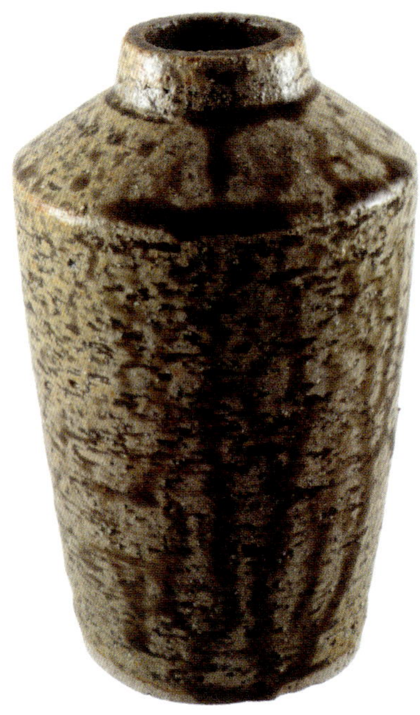

Wood ash contains all the minerals taken up by the growing tree, including calcium, silica, potassium, sodium, phosphorus, magnesium, iron and manganese. Everything else will have been burned off; most of the carbon converts to carbon dioxide. The composition of the ash will depend on the type of tree, the soil it grew in, the season the wood was cut, and whether it included bark. Soft ash, such as that from apple or beech trees, contains the least silica and the most calcium. Medium ash from oak or ash trees contains more silica. Grass ash from hay or straw contains the most silica and is classified as 'hard' ash (see table on p.92). It is also possible to use seaweed ash, which contains more sodium than wood ash. Kelp was burned in Scotland in the 18th century to produce soda ash for the glass and soap industries. When burning seaweed, make sure you are in a well-ventilated place and avoid the fumes, which can be toxic.

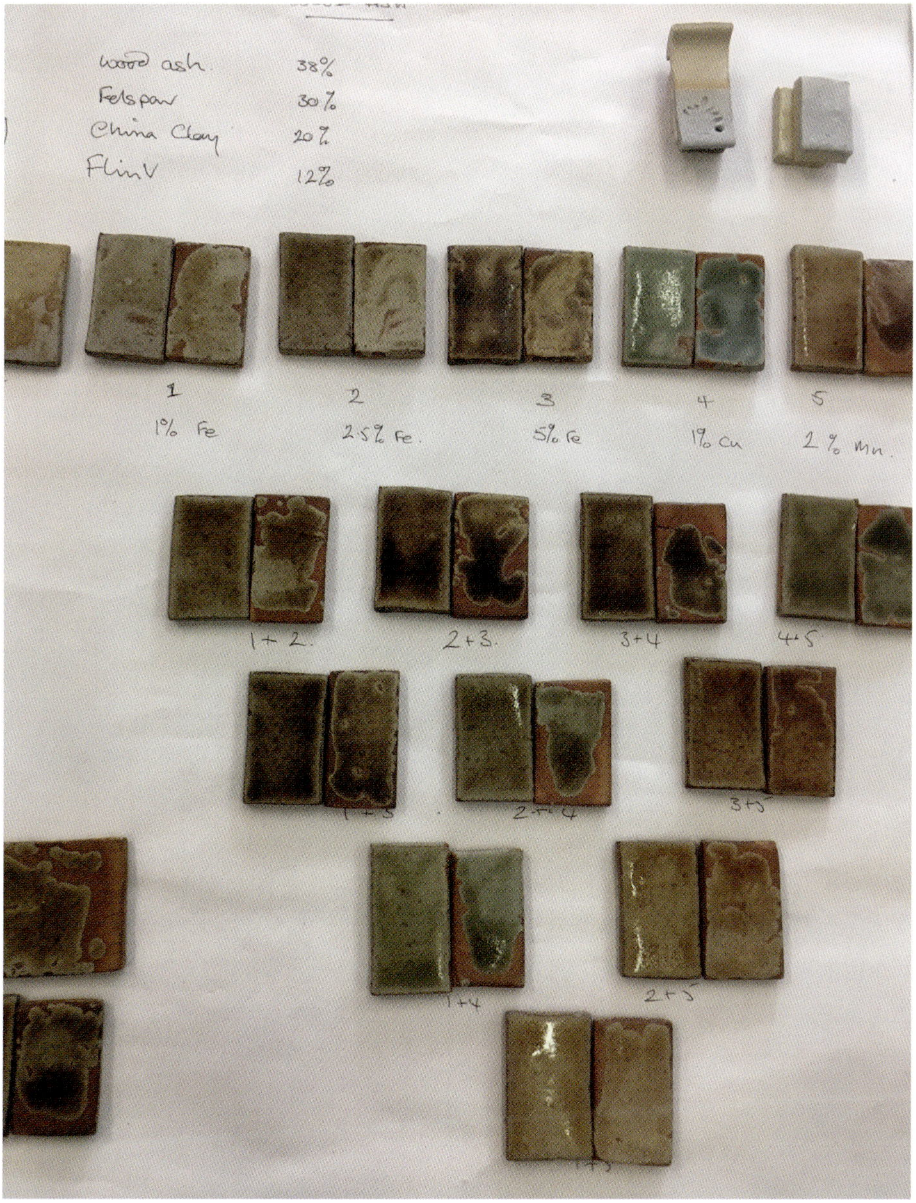

BELOW: Michael Tomalin, wood ash recipe tests showing the base glaze with additions of iron, copper and manganese.

Wood ash melts to form a glaze on its own, but it will be very runny, causing rivulets of glaze to flow down the pot. Always test ash glazes first on a vertical tile to see how much they run, and use old kiln shelves with a generous coating of batt wash (or stand tests in a specially thrown clay tray).

Wood-ash glaze, cone 8, 1250°C (2282°F).

Wood ash 38
Potash feldspar 30
China clay 20
Flint 12

Wood ash often causes crystals to form in the glaze, as it supplies excess calcium which crystallises out, particularly if the glaze is runny and low in clay and if the kiln is cooled slowly. Wood-ash glazes are often crazed, although it is possible to correct this by adding clay and silica (see transparent green ash glaze recipe on p.90). Grass ash can cause the glaze to become opaque, owing to the high silica content, particularly in thickly applied nuka glazes, which are made in Japan using rice hull ash. Ash glaze also forms naturally during wood firing, when fly ash from the burning wood settles on pots. These glaze effects are prized in Japan, where a glassy drip of molten ash glaze trickling down a pot is called a bidoro.

Richard Batterham, ash-glazed stoneware teapot.

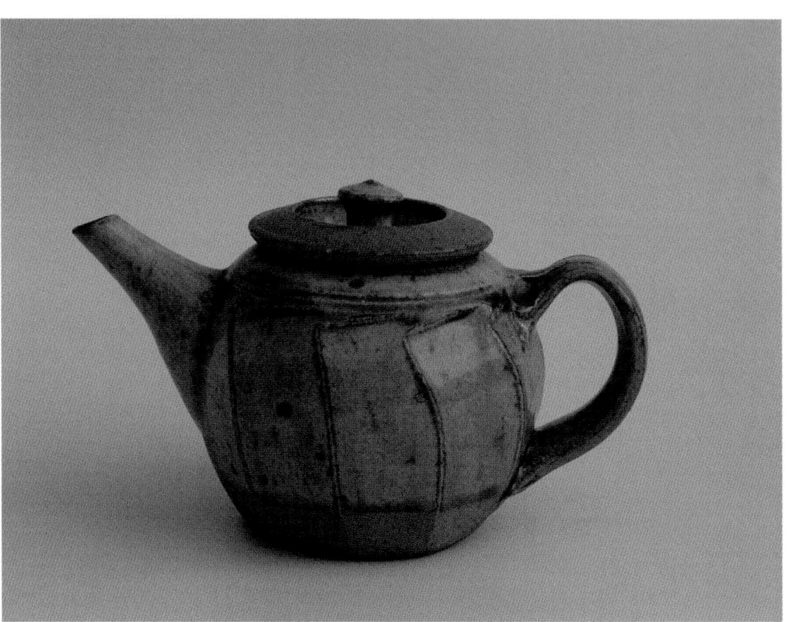

Some potters (including Katherine Pleydell-Bouverie and Norah Braden, who trained under Bernard Leach in the 1920s) have made a careful study of different ashes, keeping each type of ash separate. Others use any mixed wood ash they can get hold of, burned in the fireplace or wood burner. Care should be taken not to contaminate the ash with soil, which contains iron oxide. Likewise, coal ash is high in iron oxide and can give darker brown colours. You can use pure ash on its own, blend 50:50 with feldspar or clay or add a small amount to an existing glaze recipe. (See also ash glaze test tile images and recipes in crackle section on pp 86–7.)

Wood ash analysis (from *The Potter's Dictionary,* Hamer and Hamer, 2015)

	Ash	Apple	Beech	Oak	Larch	Pine	Spruce	Meadow grass	Wheat straw
Silica SiO_2	24	2	6	10	11	18	4	58	70
Calcia CaO	27	65	56	51	27	32	55	10	6
Potash K_2O	17	15	17	11	21	18	14	15	13
Soda Na_2O	8	6	4	6	9	6	10	4	2
Magnesia MgO	12	6	11	9	8	6	10	5	4
Phosphorus P_2O_5	7	5	5	10	8	7	7	4	5
Iron Fe_2O_3	4	1	1	1	4	4		1	
Manganese MnO				1	11	4			
Alumina Al_2O_3	1			1	1	5		3	

RIGHT: Stephen Parry, vase, wood-fired stoneware, pine-ash glaze: potash feldspar 25, pine ash 35, flint 17, Hyplas 71 ball clay 23, 25 x 7cm (9¾ x 2¾in.), 2013. *Photo: Stephen Parry.*

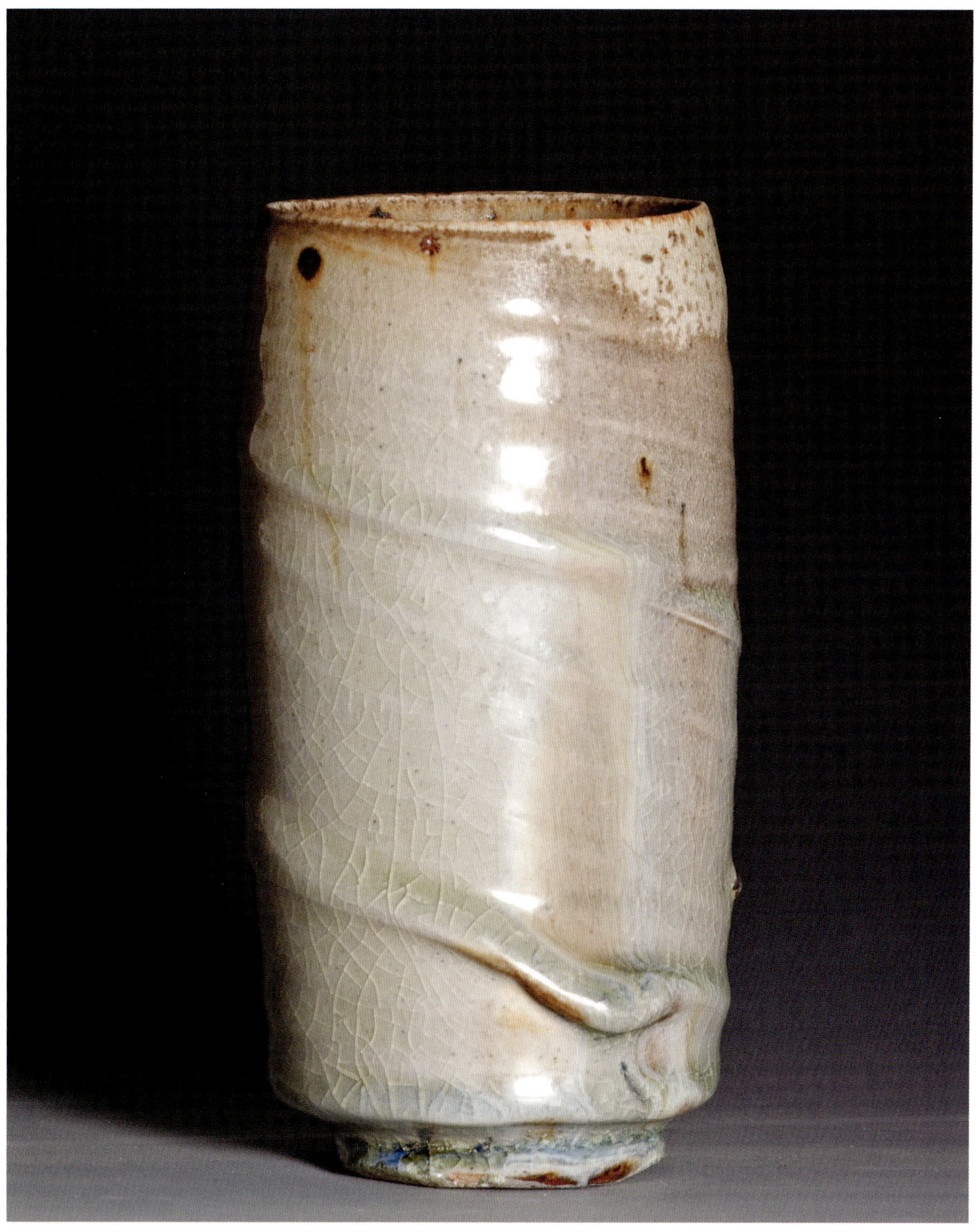

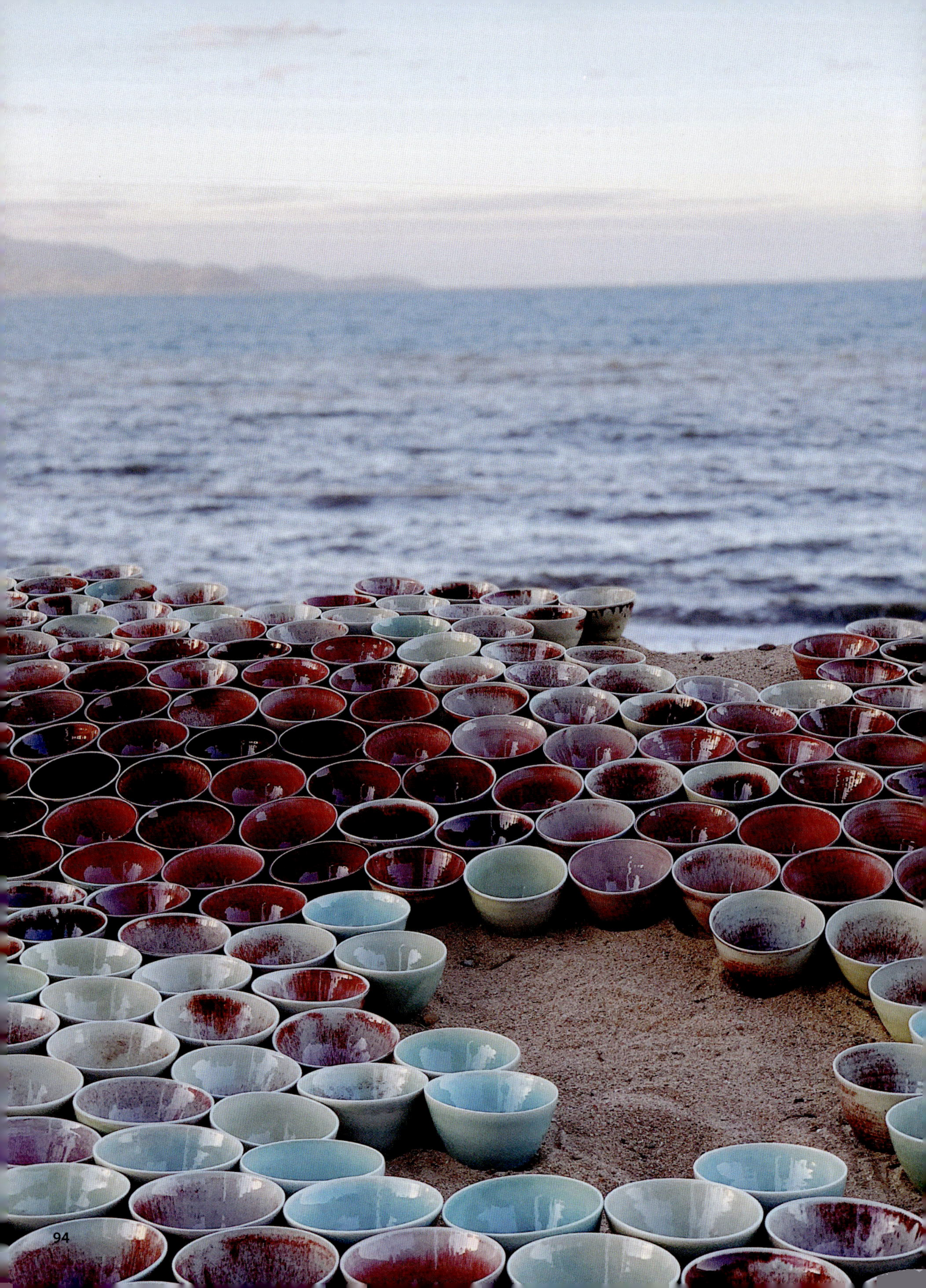

13

Celadon and Copper red glazes

Celadon glaze

Celadon glazes are transparent grey-green glazes originally made in China in the Song dynasty (960–1279 AD). The Chinese were attempting to imitate the colour of jade and made their glazes using mixtures of feldspar, clay and wood ash. Celadon glazes range from olive green to pale blue in colour and satin matt to glossy in texture. They can sometimes be crazed or crackled; covered in a fine network of cracks. The type of clay used affects the glaze colour; celadons are often green on stoneware and blue on porcelain. The blue or green colour of the celadon glaze traditionally comes from small quantities (0.5-1%) of iron oxide fired in a reduction atmosphere in a fuel-burning kiln. Any titanium dioxide impurities in the clay will combine with the iron oxide in the glaze to produce a green colour.

In this section we will explore how to create celadon-like glazes in an electric kiln. In oxidation in an electric kiln, the small amounts of iron oxide in a traditional celadon glaze fire to a straw-yellow colour. Transparent blue-green celadon-like glazes can be made in an electric kiln using a mixture of copper and cobalt oxides or a commercial blue-green stain. A less well-known method is to add a local reduction agent such as silicon carbide, silicon or aluminium powder to a traditional celadon glaze recipe to reduce the ferric iron oxide, Fe_2O_3, to ferrous iron oxide, FeO, in an electric kiln. This is explained in detail later in this section.

LEFT: Leela Chakravarti and Edward O'Brien's Strand Ephemera 2019 installation of stoneware bowls, each 12 x 16 cm (4¾ x 6¼ in.), reduction fired in a gas kiln cone 7-9 (although pockets are oxidised where the bowls are green). 'Each of our bowls represents a coral polyp and en masse they represent a giant coral colony. Transitioning from deep reds to paler pinks, speckles and pale greens, our installation represents coral bleaching. We want to spread the word about global warming and its devastating impacts on coral reefs.'

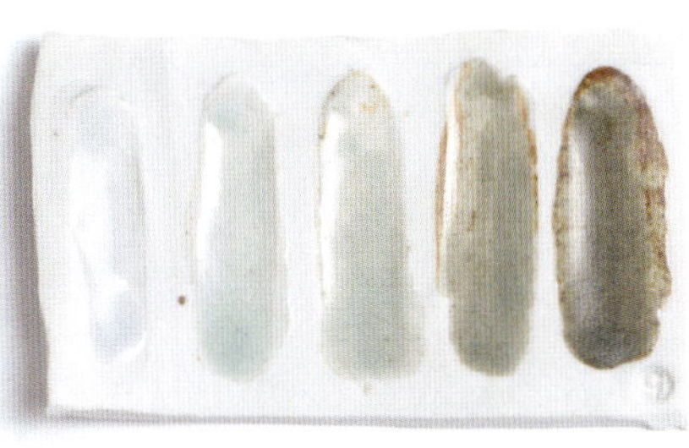

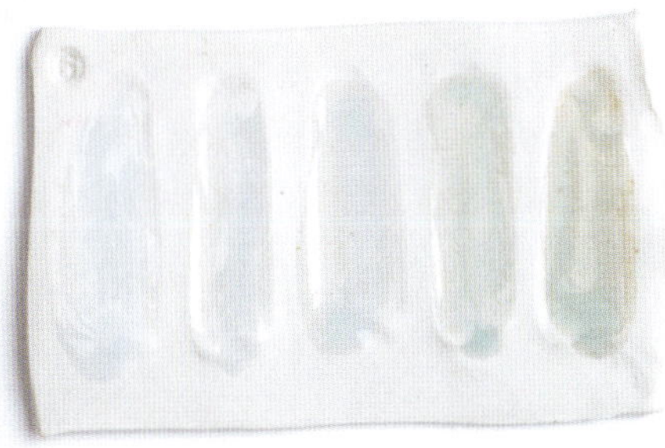

RIGHT: Celadon glaze line blends by Mirka Golden-Hann with increasing iron oxide from left to right, porcelain-fired in reduction. Two different celadon glazes both with (l–r) iron oxide + 0.5, 0.8, 1.1, 1.5, 2%.

As we have seen in the previous section, wood ash contains varying proportions of calcium, silica, potassium, sodium, phosphorus, magnesium, iron and manganese. In a reduction firing a mixture of wood ash, feldspar and clay can produce green, glassy glazes similar to celadons. To achieve this effect when firing in an electric kiln in oxidation, small amounts of copper oxide and cobalt oxide can be added to an ash glaze to make a blue-green glaze similar in colour to a celadon. This glaze may be slightly speckled unless it is sieved several times through a 100 mesh sieve. It should be applied relatively thickly but can be runny, so make sure you fire a test piece before applying it to a whole kiln load of pots.

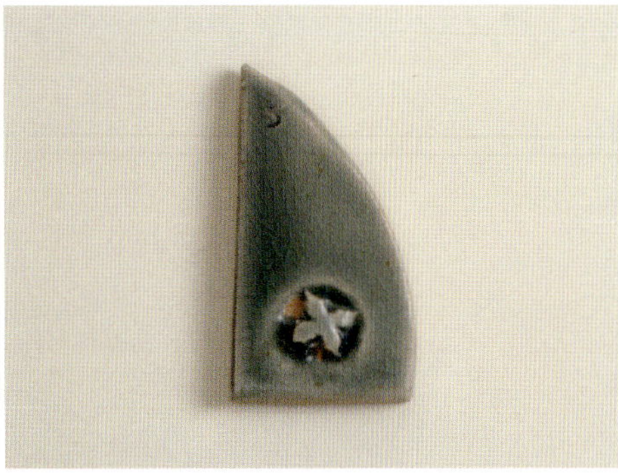

Green crackle celadon for stoneware, cone 9, 1280°C (2336°F), oxidation.

Wood ash 30
Potash feldspar 30
China clay 17
Whiting 4
Flint 21
+
Copper oxide 0.5
Cobalt carbonate 0.1

To make a reduction celadon, a small amount of iron oxide (0.5–1%) is added to a clear glaze. Blue celadons can be made if the clay and glaze materials are low in titanium dioxide, which causes the iron in the celadon to turn green in reduction. The ideal clay is porcelain, which is low in titanium and provides a white background for the celadon glaze. A blue celadon-like glaze can be made in oxidation by adding a small amount (0.1%) of copper oxide or copper carbonate instead of the iron oxide. The clay content needs to be low (less than 5%), otherwise the colour will be more green.

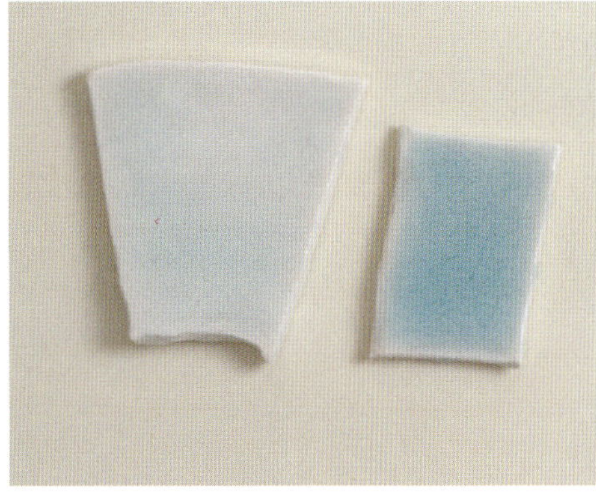

Blue celadon for porcelain, cone 6–8, 1240–1260°C (2264–2300°F), oxidation.

Soda feldspar 47
Quartz 18
Borax frit 15
Whiting 14
China clay 5
+Copper oxide 0.1 pale blue
+Copper oxide 0.5 pale turquoise

Another way to make a celadon-like glaze in an electric kiln is to use iron oxide (as you would with a traditional reduction celadon) together with a reducing agent. This will locally reduce the iron oxide in the glaze, giving a green celadon, even in an electric kiln. The most widely used reducing agent is very finely ground silicon carbide (600-1200 mesh), but this may produce bubbles of carbon dioxide which can form blisters in the glaze. Other, less widely known reducing agents include silicon and aluminium powder. You can buy 200 mesh aluminium powder from sculpture suppliers, who use it mixed with resin to make cast metal-like sculptures. Silicon is also available as a fine powder. This is the pure semi-metal used to make silicon wafers for the semiconductor industry, not to be confused with silica (silicon dioxide) or silicone (rubber). All these reducing agents will react with oxygen in the glaze during firing, causing the iron in the glaze to change from a honey-yellow colour (Fe^{3+}) to a blue-green colour (Fe^{2+}).

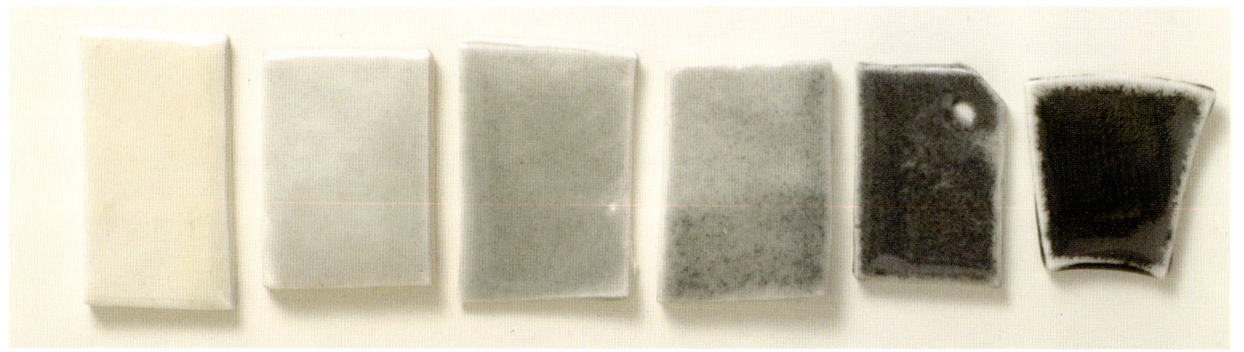

Line blend using 0.5 yellow iron oxide and 0, 0.1, 0.2, 0.3, 0.4, 0.5 aluminium powder (200 mesh), in the green celadon glaze shown below.

The idea of using aluminium powder was suggested by Dr William Carty, professor of Ceramic Engineering at Alfred University, New York, USA.

Aluminium powder can be unstable as it reacts readily with oxygen and does not always give consistent results. I have found that the glaze is very sensitive to the amount of aluminium powder used; add too much and the glaze will have black speckles of aluminium or even turn completely black, although both of these can be viewed as interesting effects. The following celadon recipe has 0.2% aluminium powder. Several different forms of iron can be used. The usual form of iron oxide used by potters is red iron oxide, Fe_2O_3 but yellow iron oxide $Fe_2O_3.H_2O$ can also be used, although is a slightly weaker form. It may be more difficult to get a blue celadon using this method, as any oxidised iron will remain yellow and, combined with the blue reduced iron, will make green.

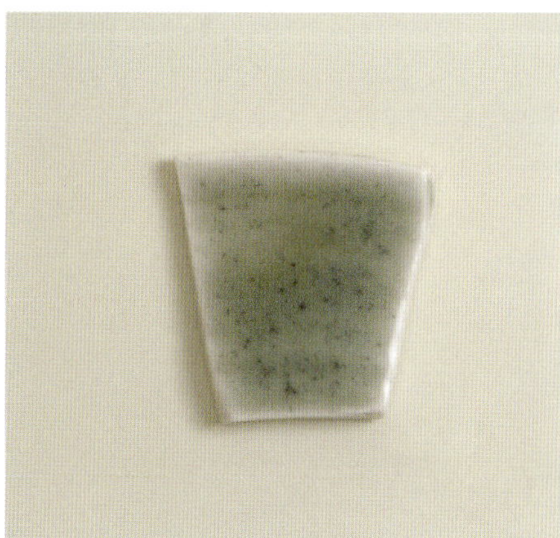

Green celadon using reducing agent, cone 6–8, 1240–1260°C (2264–2300°F), oxidation.

Potash feldspar 34
Borax frit 14
Whiting 11
China clay 13
Quartz 23
Dolomite 5
+
Yellow iron oxide 0.75
Aluminium powder 0.2

In conclusion, there are several ways to make celadon-like glazes in an electric kiln. You can colour a clear glaze using small amounts of copper oxide and cobalt oxide, or a blue-green commercial stain. However, it may take a series of tests to find the exact shade of blue-green you want and there may be specks of cobalt oxide unless you sieve through a fine-mesh sieve. The other way is to use iron oxide together with a finely ground reducing agent such as silicon carbide, silicon or aluminium powder. This will give a more authentic green celadon but reducing agents can be difficult to control and there may be black speckles still remaining in the glaze.

Copper red

Copper red glazes are usually fired in reduction in a gas kiln, but using a similar approach to the oxidation celadon glazes in the previous section, they can also be made using fine silicon carbide (600–1200 mesh) to reduce the copper oxide in an electric kiln. A stabiliser such as tin oxide (1–2%) helps to retain the red colour during firing. The glaze should be relatively runny and only a small amount of copper oxide (0.3–0.5%) is needed to get a strong oxblood red with a similar amount of silicon carbide. The silicon carbide reacts with the copper oxide (CuO) locally in contact with it and reduces it to red copper oxide, Cu_2O, and copper metal, which form red particles in the glaze, and carbon dioxide gas, which is liberated. These glazes need to be sieved well and stirred frequently to disperse the silicon carbide. You may need to flocculate the glaze by adding small amounts of Epsom salts and bentonite (see p.60).

Copper red glaze, cone 6–8, 1240–1260°C (2264–2300°F), oxidation.

Soda feldspar 47
Quartz 18
Borax frit 15
Whiting 14
China clay 5
+
Tin oxide 1
Copper oxide 0.3
Fine silicon carbide 0.3

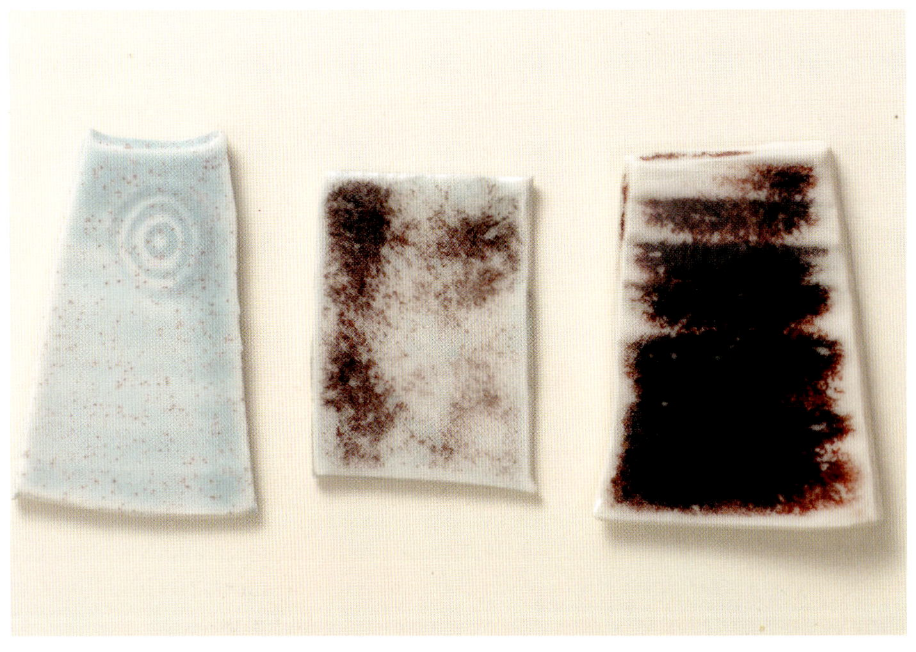

Copper red glaze tests using coarse silicon, medium silicon carbide (120 mesh) and fine silicon carbide (1200 mesh).

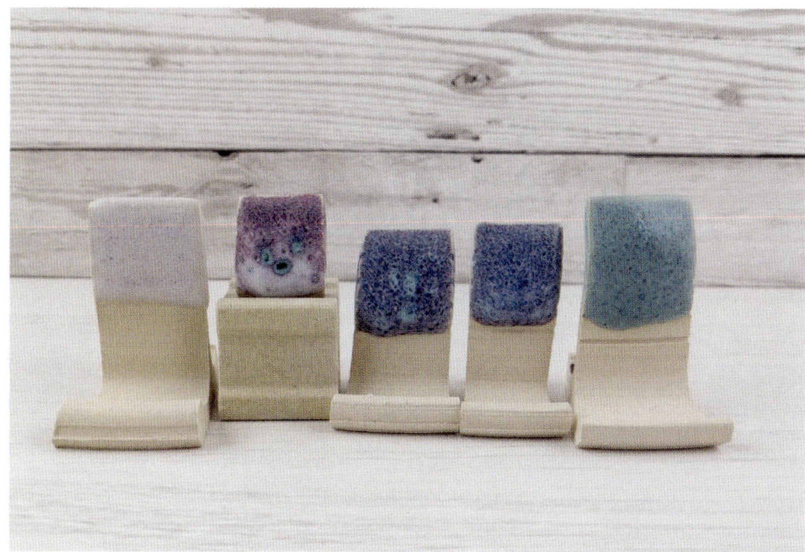

Nebula glaze (from Joe Thompson), cone 5–7.

Adapted from David Tsabar Peacock glaze, a Chun glaze reduced using silicon carbide.

Potash feldspar 46
Silica 20
Ferro frit 3134 13
Wollastonite 13
Bentonite 3
Zinc oxide 3
Bone ash 2
+
Tin oxide 2
Copper carbonate 1
Silicon carbide 1200 mesh 1

ABOVE: Joe Thompson, Nebula glaze line blend with copper carbonate 0, 0.5, 1, 1.5 and 2%.

RIGHT: Joe Thompson, mug, Nebula glaze with 1% copper carbonate (developed from David Tsabar's Peacock Glaze, see p.10).

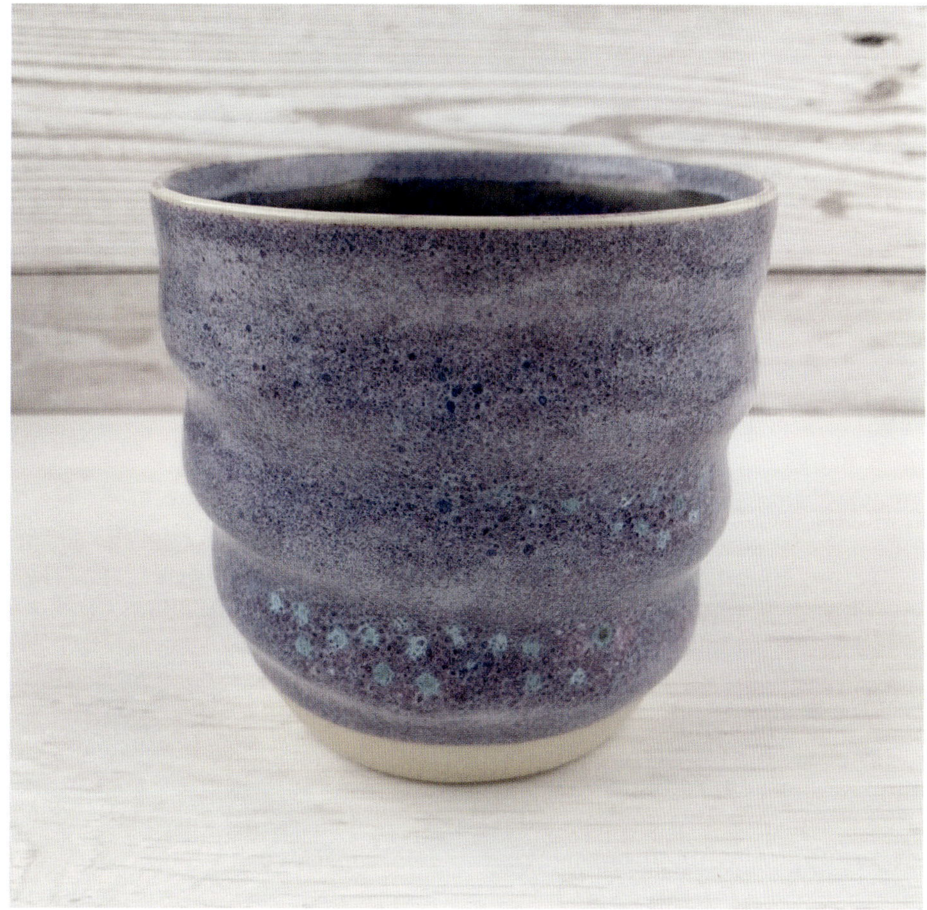

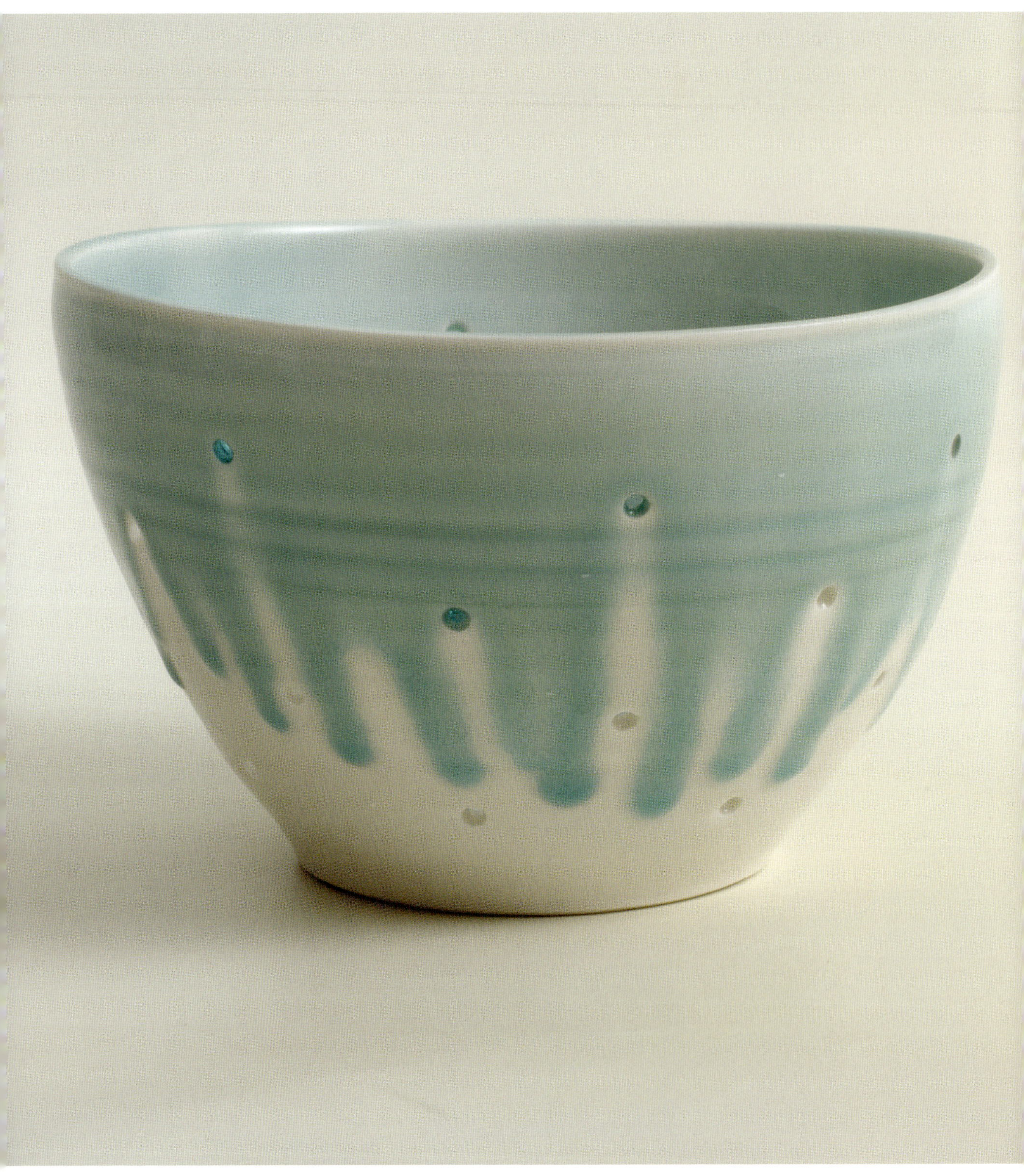

14 Drippy glazes and Chun glazes

A popular glaze effect involves drips of glaze running down a pot. Drips can be encouraged by applying glaze more thickly in certain areas. Runny glazes are made by over-firing a glaze – for example, firing a cone 6 glaze to cone 8. These runny glazes must be applied and fired very carefully to avoid the drips of glaze sticking the pot to the kiln shelf. It helps if the glaze is applied only to the top half or third of the pot. The kiln shelves should be batt-washed, or an old piece of batt-washed kiln shelf or specially made catch dish can be placed under the pot. To make your own drippy glaze, just add 5% frit to a transparent glaze and continue adding frit in 1% increments until the drips start to run at your chosen firing temperature. The extent of running depends on firing temperature, firing time, application thickness and clay body; the same glaze will run more on porcelain than on stoneware, which is more porous. To make very thickly applied drips and globs of glaze, it may be better to use a slightly more viscous glaze with a higher clay content.

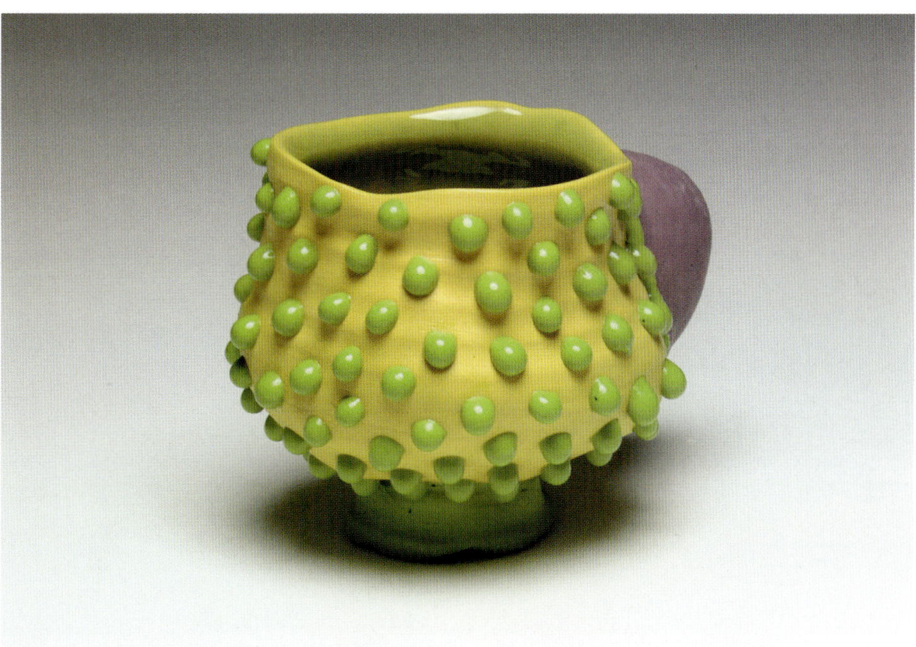

ABOVE: Nick Weddell, *Zesty Mister*, coloured porcelain and glazes, fired to cone 10, oxidation, 15 x 11.5 x 10cm (6 x 4½ x 4in.).
LEFT: Linda Bloomfield, thrown porcelain colander with runny turquoise glaze over transparent glaze, fired to cone 8.

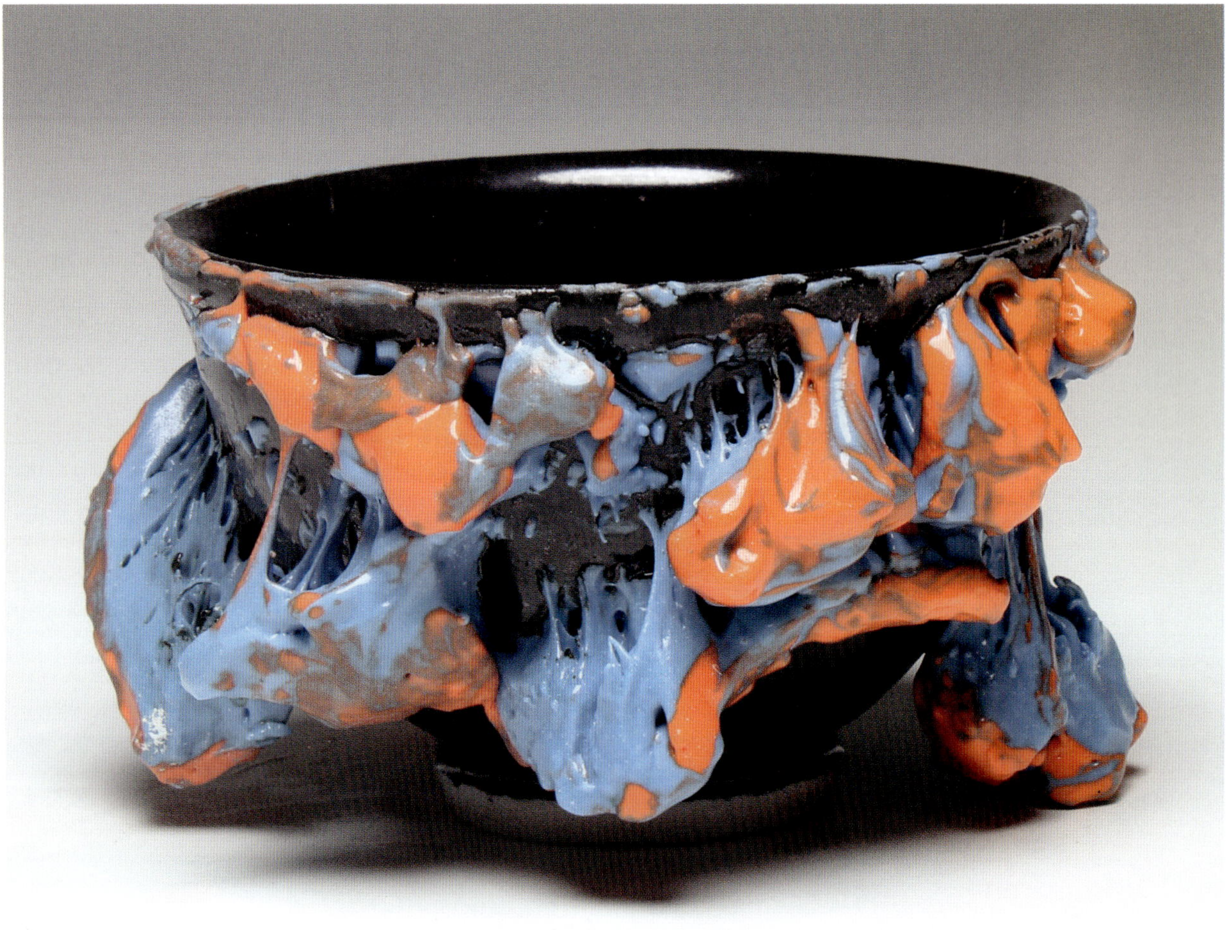

Nick Weddell, *Alien Gum Bowl*, porcelain and glazes, mixed to a thick paste with sodium silicate and water, cone 10 oxidation, 15 x 15 x 9cm (6 x 6 x 3.5in.).

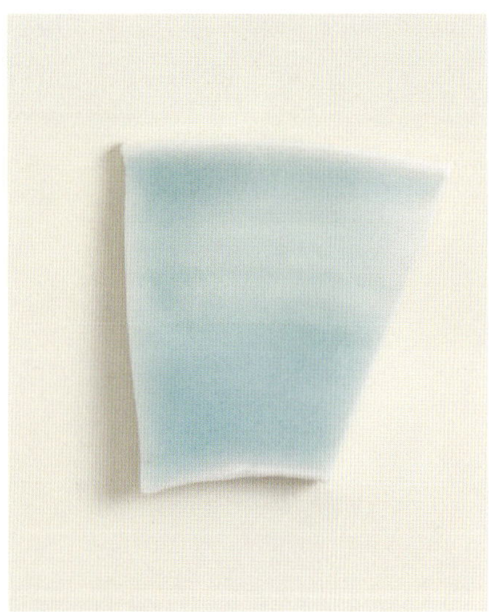

Runny turquoise glaze, cone 6–8, 1240–1260°C (2264–2300°F), oxidation.

This glaze was developed through increasing the amount of boron, by changing the frit from borax frit to calcium borate frit.

Soda feldspar 45
Quartz 17
Calcium borate frit 15
Whiting 14
China clay 5
Tin oxide 5
+
Copper oxide 1

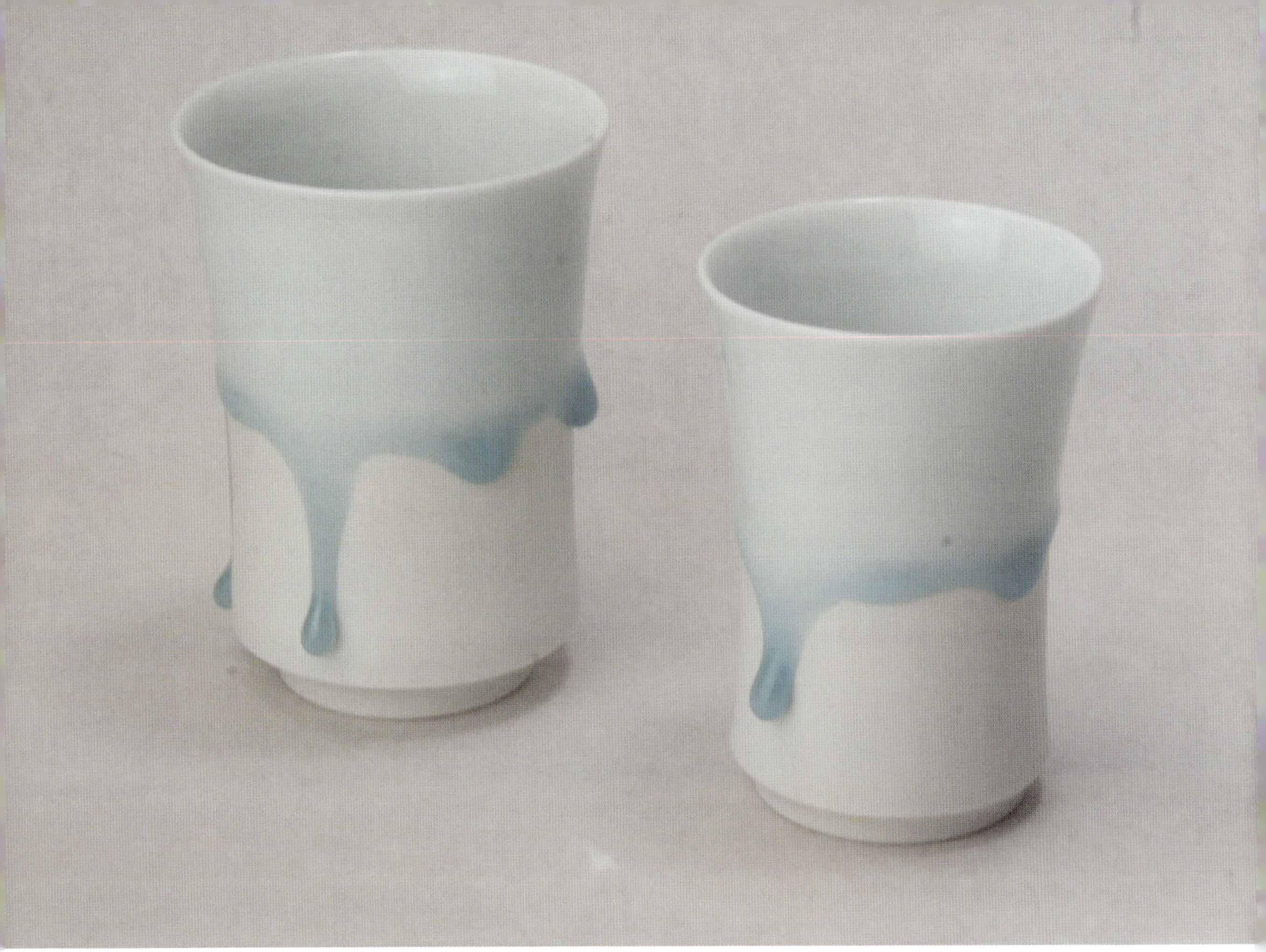

ABOVE: Nishi Takayuki, Shizuku porcelain cups with drippy blue celadon glaze, fired in reduction, Arita, Japan.

RIGHT: Alice Duck Ceramics, cobalt blue tumbler, slipcast porcelain with lines of coloured slip, drippy blue glaze, fired to 1280°C (2336°F) with a 42-minute soak, 7 x 8cm (2¾ x 3¼in.).

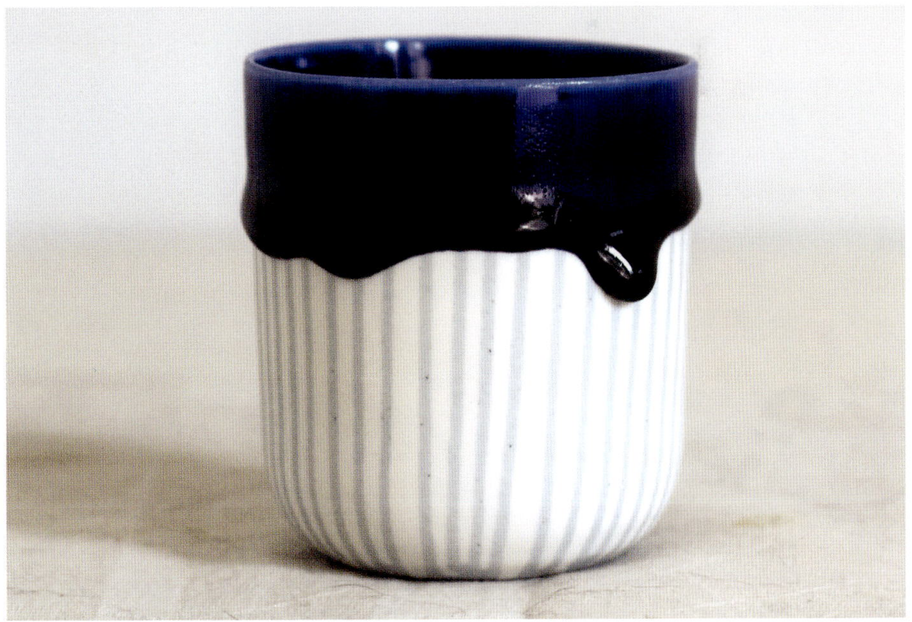

Chun glazes

Opalescent, drippy Chun (or Jun) glazes were first developed in northern China during the Song dynasty (960–1279 AD). The glazes were thick, opaque pale blue, sometimes with purple splashes from reduced copper oxide. It is thought that the opaque blue colour comes from light scattered by tiny globules of phosphorus-rich glass suspended in the surrounding silica glass. If you think of the phosphorus and silica glasses like this, they form two distinct, separate phases, like bubbles dispersed in a liquid. The globules are very small colloidal particles, a few hundred nanometres in diameter, dispersed through the transparent glaze. Short-wavelength blue light is scattered more strongly than other colours of light by the globules. This 'optical' effect gives a blue tinge to the glaze. The more globules and the bigger they are, the stronger the scattering and the cloudier the glaze appears. The glaze must be applied thickly and should contain some phosphorus and iron oxide (see cup by Joanna Howells). The phosphorus can come from wood ash or bone ash added to the glaze. Calcium borate or colemanite gives a similar effect. This type of glaze works best in reduction on a stoneware clay body or layered over a dark, iron-rich glaze such as tenmoku.

There are several ways to obtain a 'Chun' effect in glazes. Phosphorus is present in wood ash, bone ash and synthetic calcium phosphate, all of which can be added to the glaze. Calcium borate frit or colemanite can produce a similar cloudy opalescence in glazes. Rutile and titanium can also cause a similar effect in runny glazes. The 'Chun' effect causes streaking on vertical surfaces, while mottling occurs more often on horizontal surfaces of pots. Chun-type glazes are often high in silica, low in clay and very runny. They should be applied thickly at the top of the pot and more thinly lower down to avoid drips sticking to the kiln shelf.

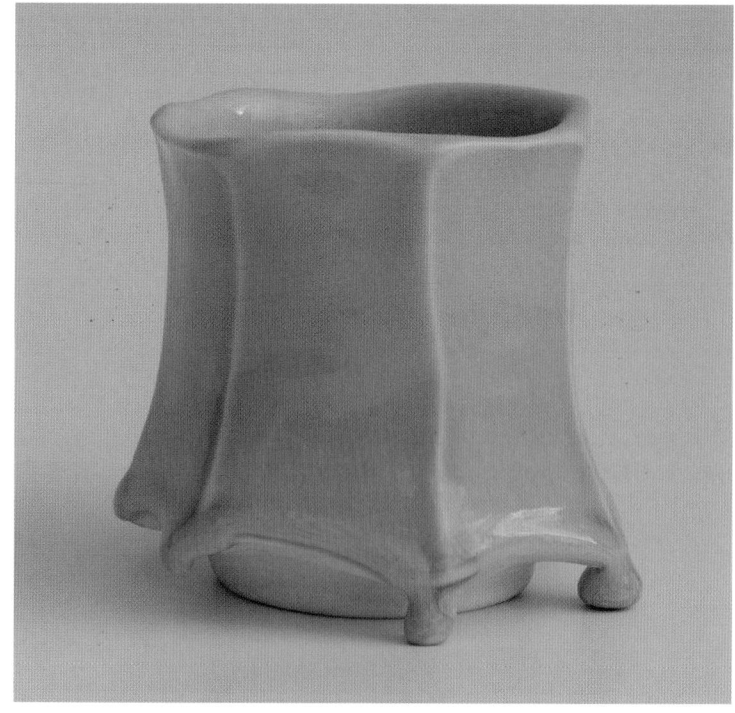

Joanna Howells, faceted porcelain cup, drippy Chun glaze, fired in reduction, height: 10cm (4in.), collection of the author.

Pale blue Chun glaze (Derek Emms), cone 9, 1280°C (2336°F), reduction.

Potash feldspar 40
Quartz 30
Whiting 20
Calcium borate frit 10
Talc 5
China clay 2
+
Black iron oxide 1

Chun glazes using ash

Some types of ash glaze also show streaking, mottling and opalescence. Wood ash varies in composition, but it often contains calcium, potassium, sodium, magnesium, phosphorus and iron oxide. Simple ash glazes can be made from wood ash and clay, sometimes with added feldspar (see recipes and tiles showing crackle ash glazes pp 86–7). A Japanese glaze called nuka is made from rice hull ash, which is high in silica. Grass, straw and reed ashes are also rich in silica and can give similar effects. Ash glazes are usually fired in reduction at cone 10 to bring out the subtle shades of green and blue derived from reduced iron oxide. As well as streaking, there may be some crystallization. This occurs when the kiln is cooled slowly and crystals start to form in the molten glaze. The crystals are often formed from calcium silicate or magnesium silicate, which grow small, round crystals in a matrix of glassy, transparent glaze (see next section on crystalline glazes).

Calcium borate glazes

Calcium borate or colemanite (or Ferro frit 3134) can produce cloudy opalescence in low alumina glazes. If small crystals of calcium borate form, they can remain suspended in the surrounding silica glass and scatter light, giving an opaque or cloudy appearance. This can happen when the glaze is applied thickly or if the molten glaze pools at the bottom of a bowl. It seems to occur more reliably if the glaze has been kept for a long time in the bucket. This may be because the frit sinks to the bottom between stirrings and the calcium borate content of the glaze increases over time as the glaze is used. However, if the boron content becomes too high and the molten glaze pools very thickly, bubbles and blisters may form which must be ground down and re-fired.

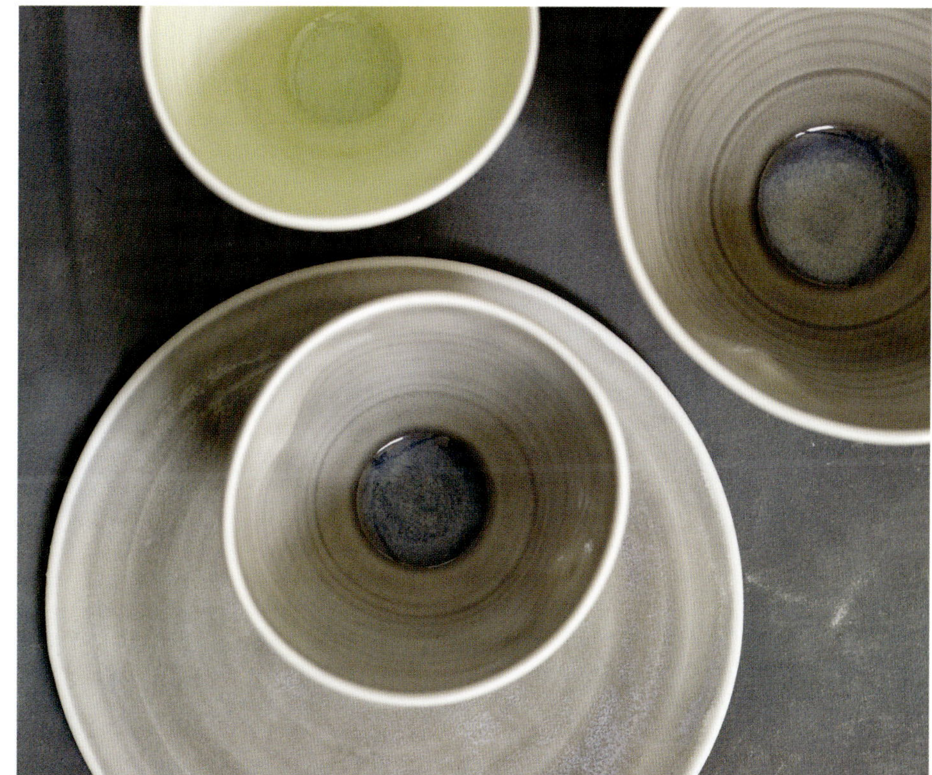

Linda Bloomfield, porcelain bowls with grey glaze containing calcium borate frit. The bluish opalescence shows at the bottom of the bowls where the glaze is thick. This only happens with calcium borate when using the last dregs from the bottom of the glaze bucket, but adding 5% rutile can increase the effect. The yellow glaze contains praseodymium oxide. *Photo: Emma Lee.*

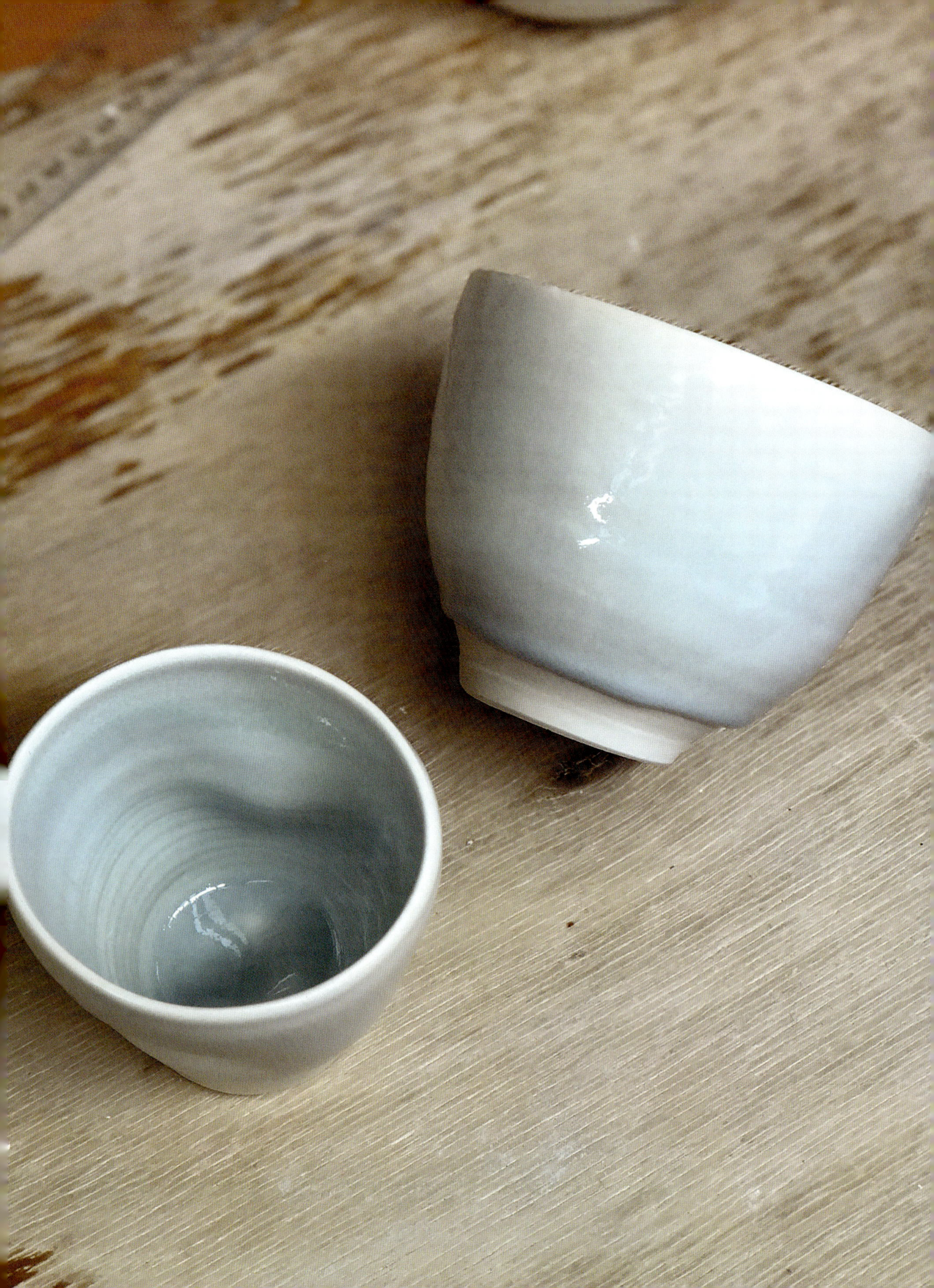

A Chun-like turquoise-blue glaze can be made by firing a copper red glaze in oxidation. A small amount of copper oxide (0.1–0.5%) and 5% tin oxide gives an opaque pale blue in oxidation, similar in colour to a reduction Chun. In this glaze, the opacity comes from the tin oxide. Opalescent glazes are not confined to blue. Other colours are possible using different colouring oxides such as the rare earth oxides, which include praseodymium, neodymium and erbium oxides. See. p.99 for a reduction Chun using silicon carbide.

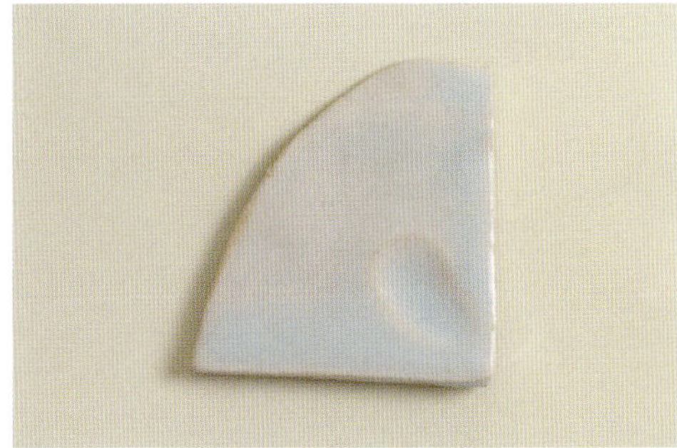

Pale copper blue glaze, cone 8, 1250°C (2282°F), oxidation.

Soda feldspar 45
Quartz 17
Calcium borate frit 15
Whiting 14
China clay 5
+
Tin oxide 5
Copper oxide 0.1

Rutile and titanium glazes

Streaking and mottling can occur in low-alumina glazes containing rutile and iron oxide. This type of glaze is sometimes called floating blue and is best over a dark iron-bearing stoneware clay body, fired in reduction. Similar effects can be obtained in oxidation with either rutile or titanium dioxide. Rutile and titanium dioxide (1–5%) cause opacity in the glaze by forming tiny crystals. In oxidation, rutile blues may need the addition of some cobalt oxide or cobalt carbonate. When both rutile and calcium borate are present in a glaze, the Chun effect is more likely to occur.

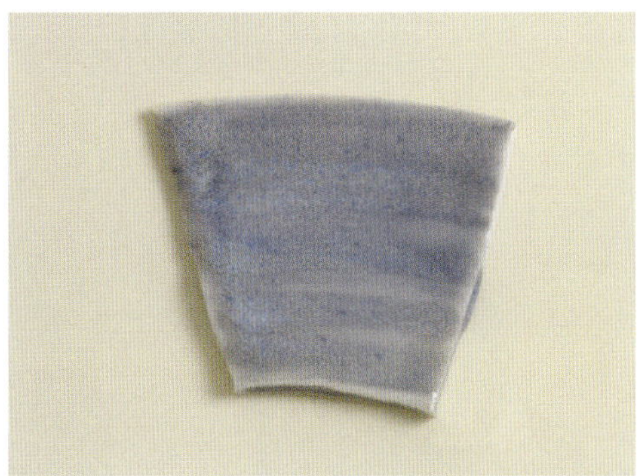

Rutile blue glaze, cone 8, 1250°C (2282°F), oxidation.

Potash feldspar 34
Calcium borate frit 14
Whiting 11
Dolomite 5
China clay 13
Quartz 23
+
Cobalt carbonate 0.4
Rutile 5

In summary, opalescent, Chun-like glazes can be made using a low-alumina glaze with the addition of:

1. Phosphorus in the form of bone ash, wood ash or grass ash.
2. Boron from calcium borate frit or colemanite.
3. Titanium or rutile.

Ash glazes and traditional Chun glazes work best in reduction. In oxidation, adding both boron and rutile to a glaze gives a reliable Chun effect.

LEFT: Linda Bloomfield, porcelain beaker and bowl, copper blue glaze fired to cone 8 in an electric kiln.

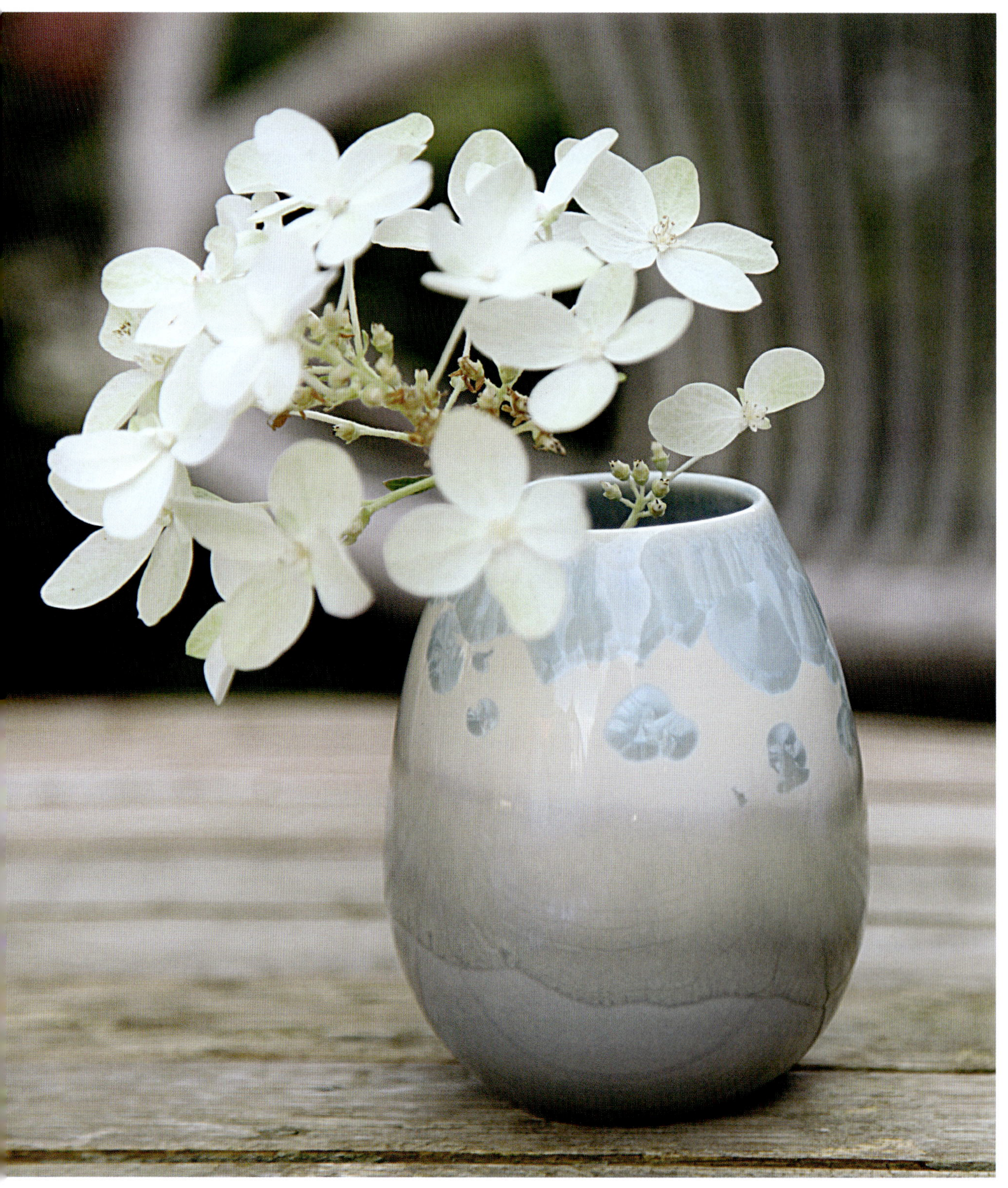

15

Crystalline glazes

Crystalline glazes can be made by adding titanium dioxide or rutile to a glaze which is relatively low in clay (with less than 5% clay). On the Stull map below, crystalline glazes are shown to form in the range 0.05–0.4 molecules alumina in the unity molecular formula. This is a wider range than for zinc silicate macro-crystalline glazes, which contain very little alumina. Too much alumina in the glaze tends to suppress crystal growth, although it is possible to make micro-crystalline matt glazes with a higher clay content. Small, round crystals form when calcium and magnesium are added to the glaze in the form of dolomite, particularly when zinc oxide is also added. Larger crystals grow in zinc silicate glazes, which are cooled very slowly by holding them at 1060°C (1940°F) for several hours at the end of firing. As these glazes are low in clay, they can be very runny, so care should be taken not to apply them too thickly on the outside of pots.

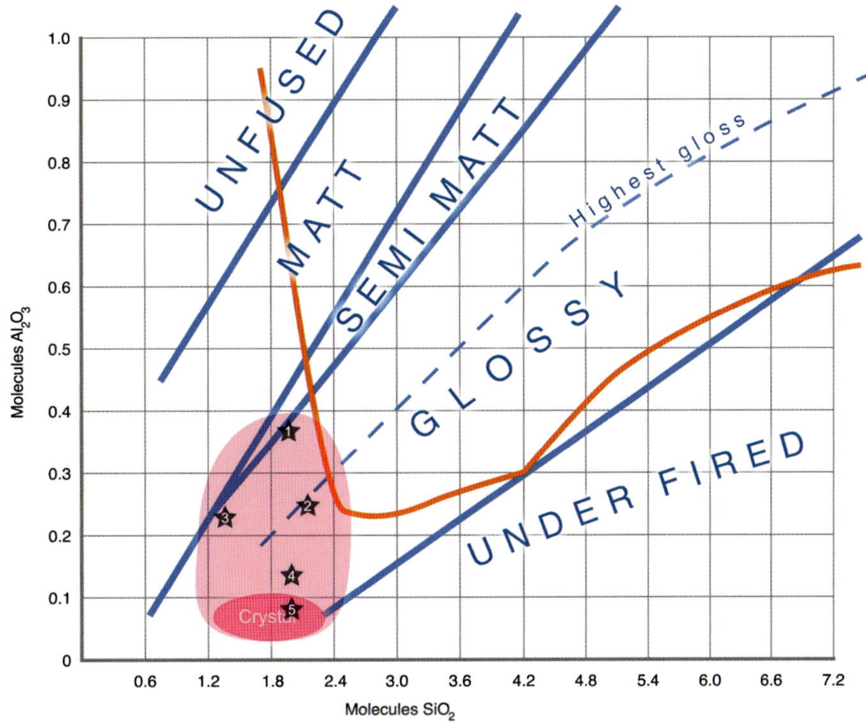

Graph of alumina and silica in porcelain glazes fired to cone 11 with constant flux 0.3 K_2O and 0.7 CaO. The ratio of 1:5 alumina to silica gives a semi-matt glaze, while 1:8 gives a shiny glaze. The straight lines on the chart represent alumina:silica ratios of 1:4 (matt), 1:5 (semi-matt) and 1:12 (shiny, crazed glaze). The dashed line is 1:8 Al_2O_3:SiO_2 (bright, shiny glaze). Data from R.T. Stull 1912. Graphic by Henry Bloomfield.

Crystalline glazes are found in the pale pink area in the bottom left-hand corner. Macro-crystalline glazes are in the dark pink area. The numbers refer to the glaze recipes on pp 112–14.

OPPOSITE: Wauw Design, porcelain vase with overlapping crystalline and grey glazes. Made in Copenhagen 2016.

Let us look at why crystals often form in glazes which are low in alumina and contain excess calcia or magnesia. The lack of alumina in the glaze means that it is very fluid in the melt, and atoms can move around easily. Only a limited amount of calcium can dissolve in the glaze. The silica in the glaze reacts with the excess calcium to form calcium silicate crystals, which grow in the molten glaze if it is cooled slowly, giving time for the atoms to arrange themselves in a crystal structure. The calcium silicate molecules first aggregate into chains, then double chains, then sheet structures and finally three-dimensional framework structures such as anorthite (calcium feldspar). The crystals can include wollastonite (calcium silicate $CaSiO_3$), diopside ($CaMgSi_2O_6$) and enstatite (magnesium silicate $Mg_2Si_2O_6$). The latter two are types of pyroxene, a chain silicate. The glaze becomes devitrified and is no longer transparent. There can be a few crystals floating in a matrix of glossy glaze, or the crystals can cover the entire surface to form a matt glaze if the kiln is cooled slowly. Barium, strontium and zinc will also form crystals in a similar way.

OPPOSITE TOP LEFT: Molybdenum crystal glaze magnified, a combination of glazes from Herbert Sanders and Fara Shimbo, fired to cone 9, made by Avril Farley. Herbert Sanders, glaze cone 9, feldspar 39, whiting 7, barium carbonate 2, zinc oxide 7, calcium borate 17, silica 22, + molybdenum oxide 4, titanium oxide 8.

OPPOSITE RIGHT: Avril Farley, detail of nickel crystalline glaze over porcelain slip with red stain, fired to cone 9.

OPPOSITE BOTTOM LEFT: John Stroomer, zinc silicate crystalline glaze using copper and cobalt oxides.

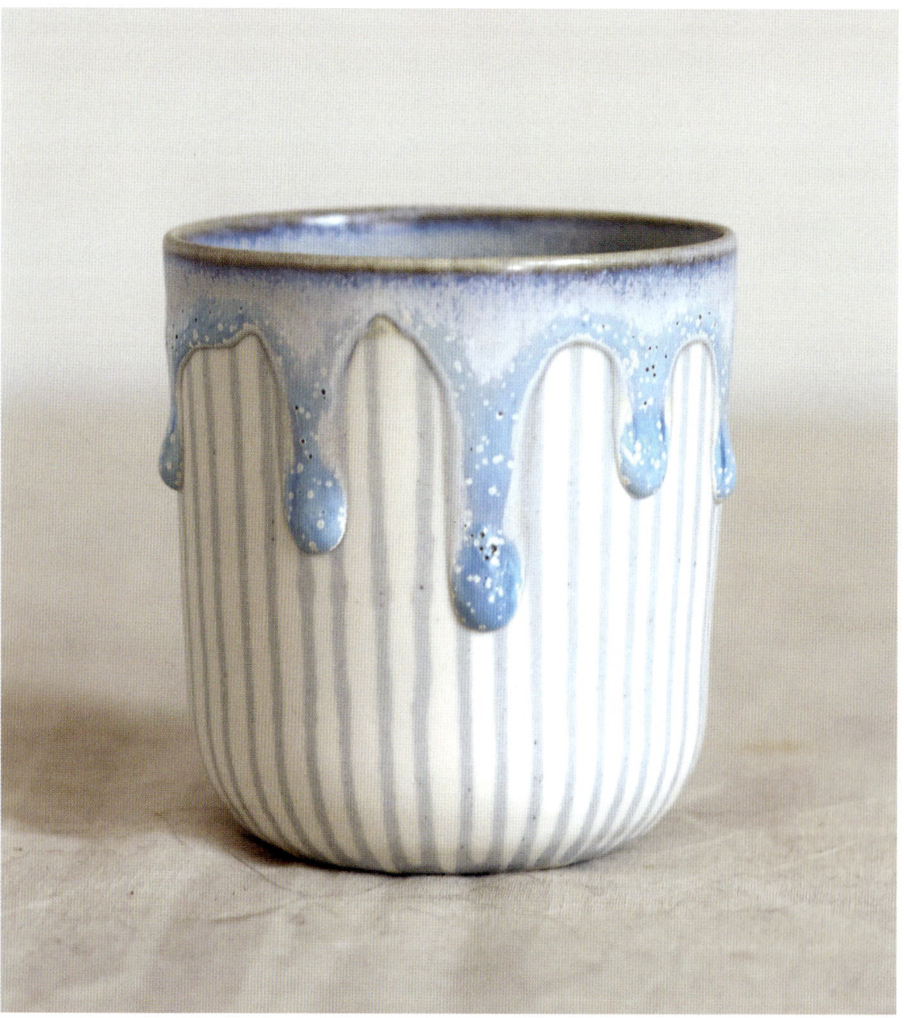

Alice Duck Ceramics, blue tumbler, slipcast porcelain with lines of coloured slip, drippy crystalline matt glaze, fired to 1280°C (2336°F) with a 42-minute soak. 7 x 8cm (2½ x 3in.).

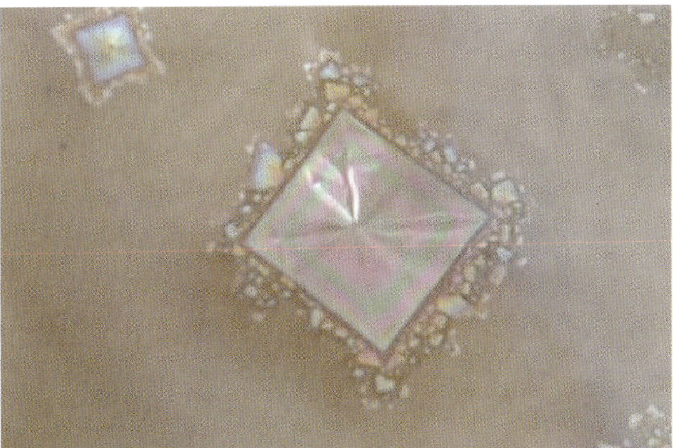

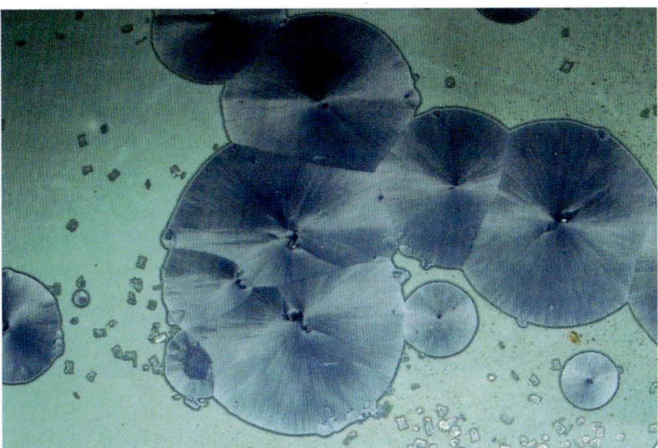

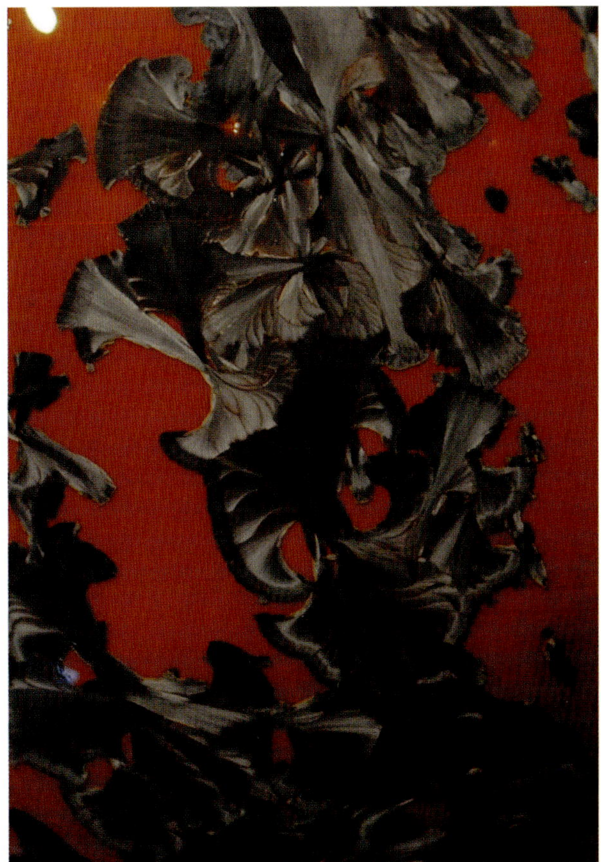

Macro-crystalline glazes with very large crystals can be made from zinc silicate if held for several hours during cooling at around 1050–1100°C (1922–2012°F) to allow time for the crystals to grow while the glaze is still molten. The zinc silicate mineral Zn_2SiO_4 is called willemite. Depending on the holding temperature, its crystals can be acicular, needle-shaped or spherulite, star-shaped and can be coloured with cobalt, copper or nickel. The crystals selectively take up certain colouring oxides in preference to others. For example, cobalt will colour the crystals blue, nickel will give steel blue and manganese will give pink, though only if there is no cobalt or nickel present. Manganese and copper oxide will usually colour the background if there are other colouring oxides present. The rare earth oxides erbium, neodymium and praseodymium can be used to colour crystal glazes. Molybdenum and tungsten are used to get iridescent metallic crystals. Titanium and rutile are used to seed crystals. In a molten glaze, they initially form zinc titanate $ZnTiO_3$ or calcium titanate $CaTiO_3$ from which other crystals can grow on cooling. At stoneware temperatures, mullite in the glaze-body interface will act as a seed for crystals, as will the addition of wood ash or bone ash to the glaze.Aventurine glazes are glittery glazes with tiny sparkling crystals containing either iron oxide (for gold-brown), chromium oxide (green) or uranium oxide (yellow-orange).

Recipe numbers correspond to numbers marked on Stull map on p.109.

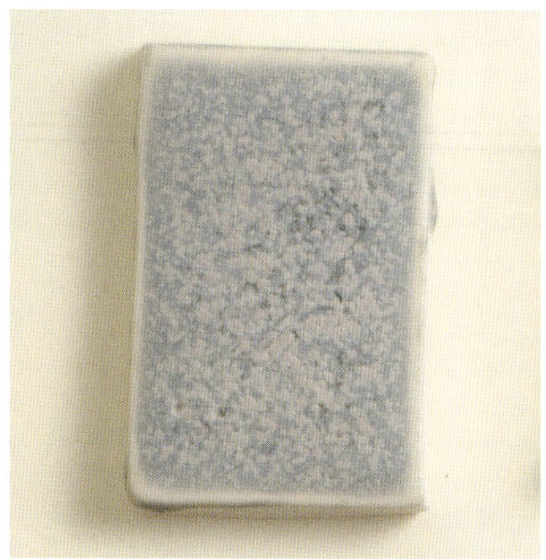

1 Matt crystalline glaze, cone 8, 1260°C (2300°F). Glaze 3 on the Stull map.

Soda feldspar 41
Dolomite 22
Quartz 11
China clay 18
Whiting 3
Zinc oxide 5
+
Cobalt oxide 0.3
Nickel oxide 0.9

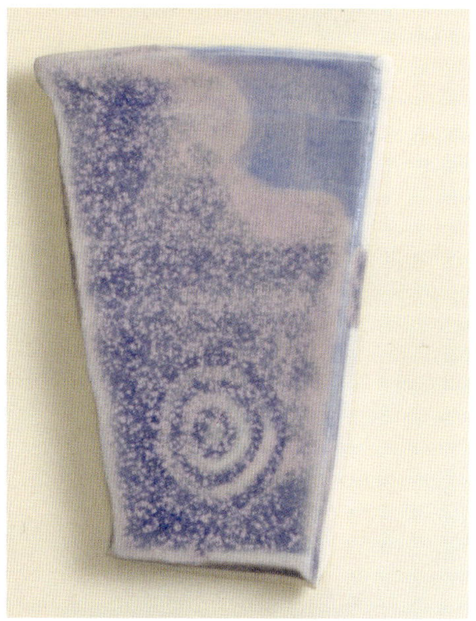

Blue and pink crystalline matt glaze, cone 8, 1260°C (2300°F).

Soda feldspar 41
Dolomite 22
Quartz 11
China clay 18
Whiting 3
Zinc oxide 5
+
Tin oxide 4
Cobalt oxide 0.75

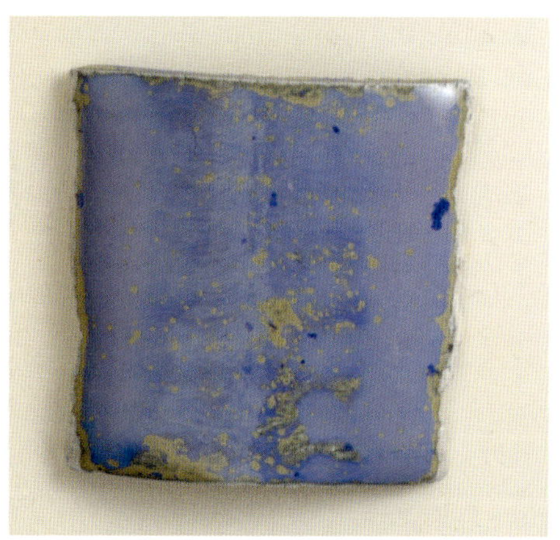

3 Semi-glossy blue and yellow crystal glaze (Lasse Östman) cone 8, 1260°C (2300°F), 30-minute soak. Glaze 3 on the Stull map.

This glaze has small, round crystals and is very runny.
Potash feldspar 63
Dolomite 16
Zinc oxide 17
Rutile 3
+
Cobalt oxide 1

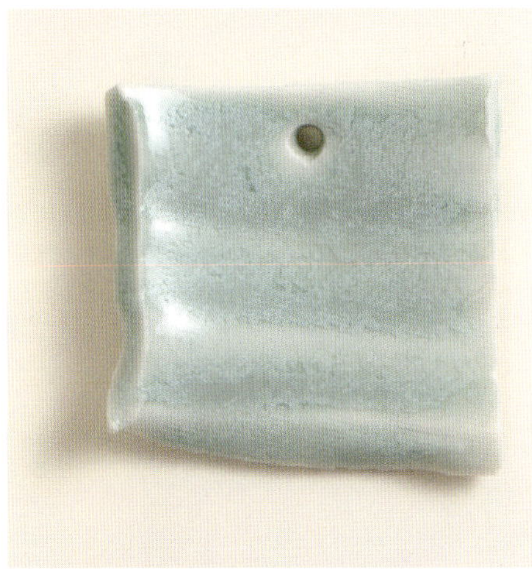

2 Magnesium micro-crystalline matt glaze, cone 6–8, 1240–1260°C (2264–2300°F). Glaze 2 on the Stull map.

Various colouring oxides can be added, such as cobalt or copper oxide. The crystals are often a different colour to the background. Matt at cone 6, more glossy and runny at cone 8.

Soda feldspar 42
Dolomite 22
Quartz 22
China clay 6
Whiting 3
Zinc oxide 5
+
Copper oxide 1

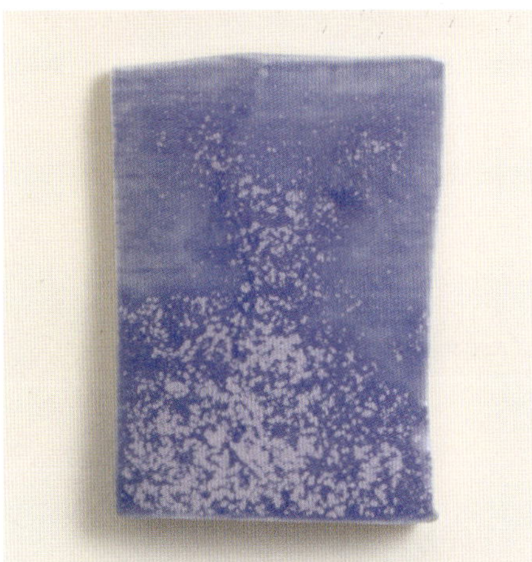

Glossy blue with pink crystals, cone 6–8, 1240–1260°C (2264–2300°F).

Soda feldspar 42
Dolomite 22
Quartz 22
China clay 6
Whiting 3
Zinc oxide 5
+
Tin oxide 4
Cobalt oxide 0.75

4 Semi-matt crystal glaze (Lasse Östman) cone 8, 1260°C (2300°F), 45-minute soak. Glaze 4 on the Stull map.

Glossy background with matt crystals.

Potash feldspar 28
Quartz 32
Zinc oxide 19
Dolomite 3
Strontium carbonate 3
Lithium carbonate 7
China clay 3
Titanium dioxide 4
+
Nickel oxide 0.5

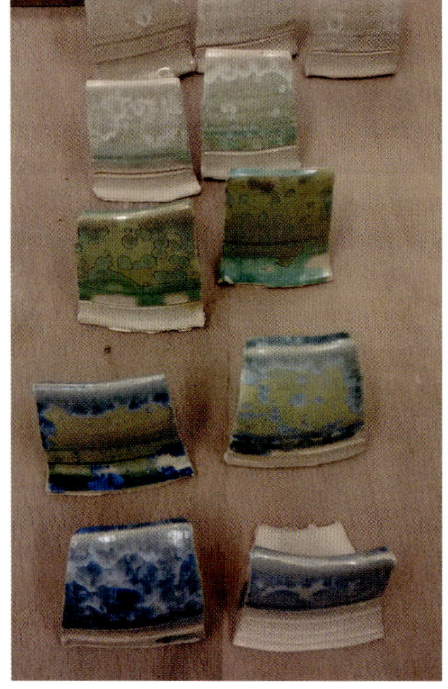

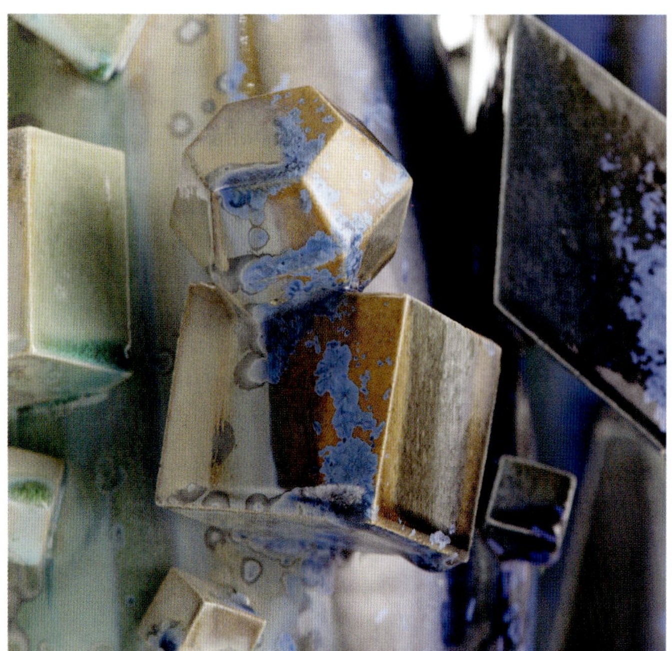

5 Macro-crystalline glaze, Avril Farley, cone 8, 1260°C (2300°F). Glaze 5 on the Stull map.

This glaze has large crystals if cooled slowly below 1100°C (2012°F).

Ferro frit 3110 47
Calcined zinc oxide 23
Calcined china clay 3
Quartz 23
Titanium dioxide 4

Blue-green aventurine glaze, cone 5–7. Not food-safe (derived from a recipe in *Clay Times* Mar/Apr 07).

Potash feldspar 17
Quartz 32
Whiting 15
Strontium carbonate 13
Borax frit 15
China clay 7
+
Copper oxide 2.5
Chromium oxide 1.5

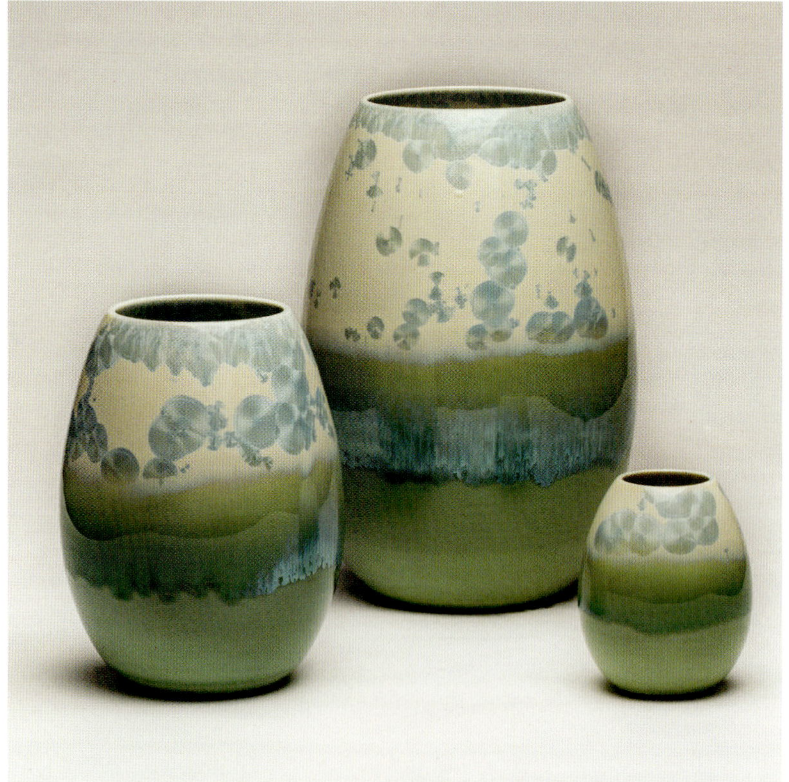

ABOVE: Wauw Design, porcelain vases with overlapping crystalline and green glazes. Copenhagen, 2016.

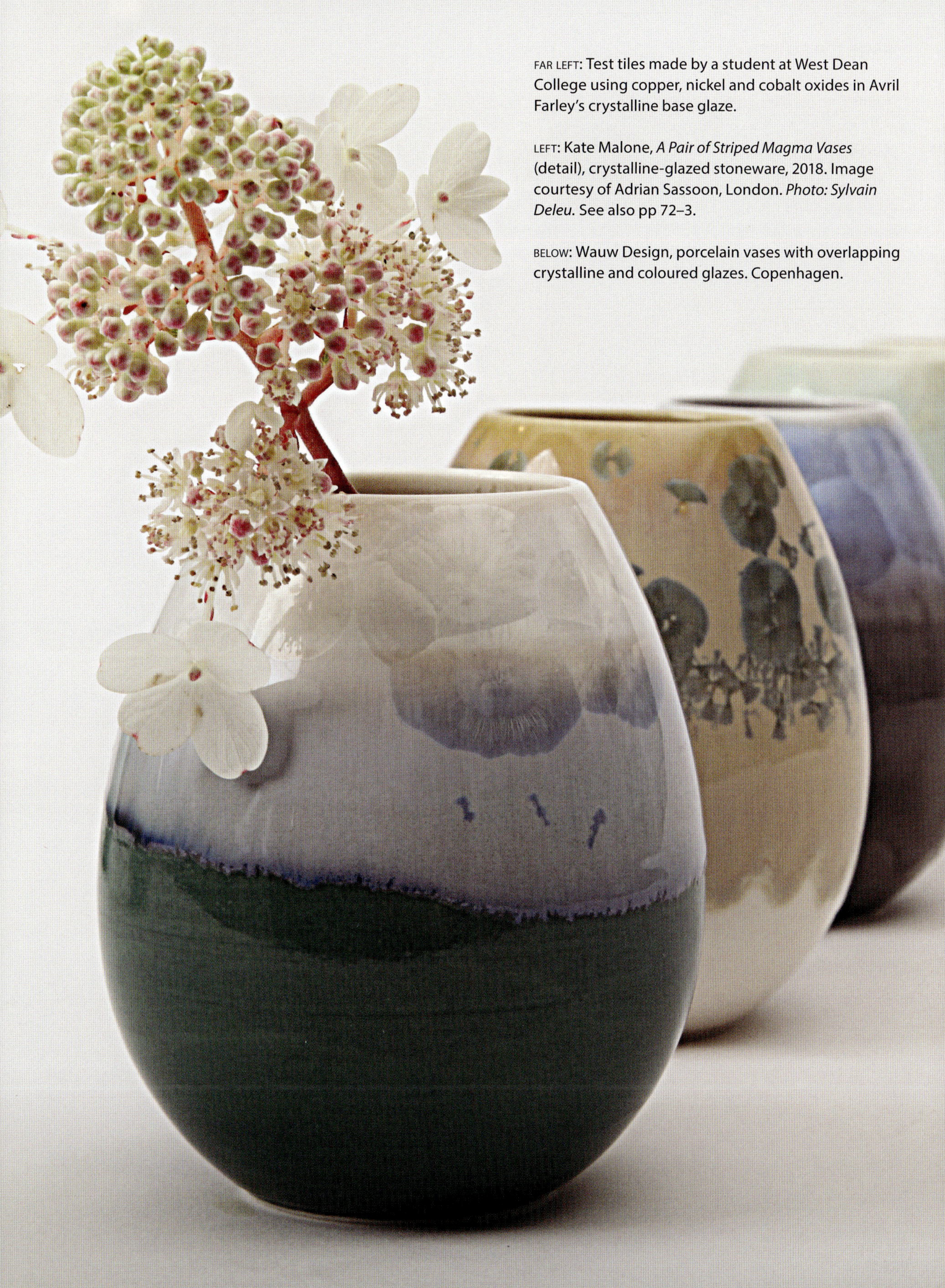

FAR LEFT: Test tiles made by a student at West Dean College using copper, nickel and cobalt oxides in Avril Farley's crystalline base glaze.

LEFT: Kate Malone, *A Pair of Striped Magma Vases* (detail), crystalline-glazed stoneware, 2018. Image courtesy of Adrian Sassoon, London. *Photo: Sylvain Deleu*. See also pp 72–3.

BELOW: Wauw Design, porcelain vases with overlapping crystalline and coloured glazes. Copenhagen.

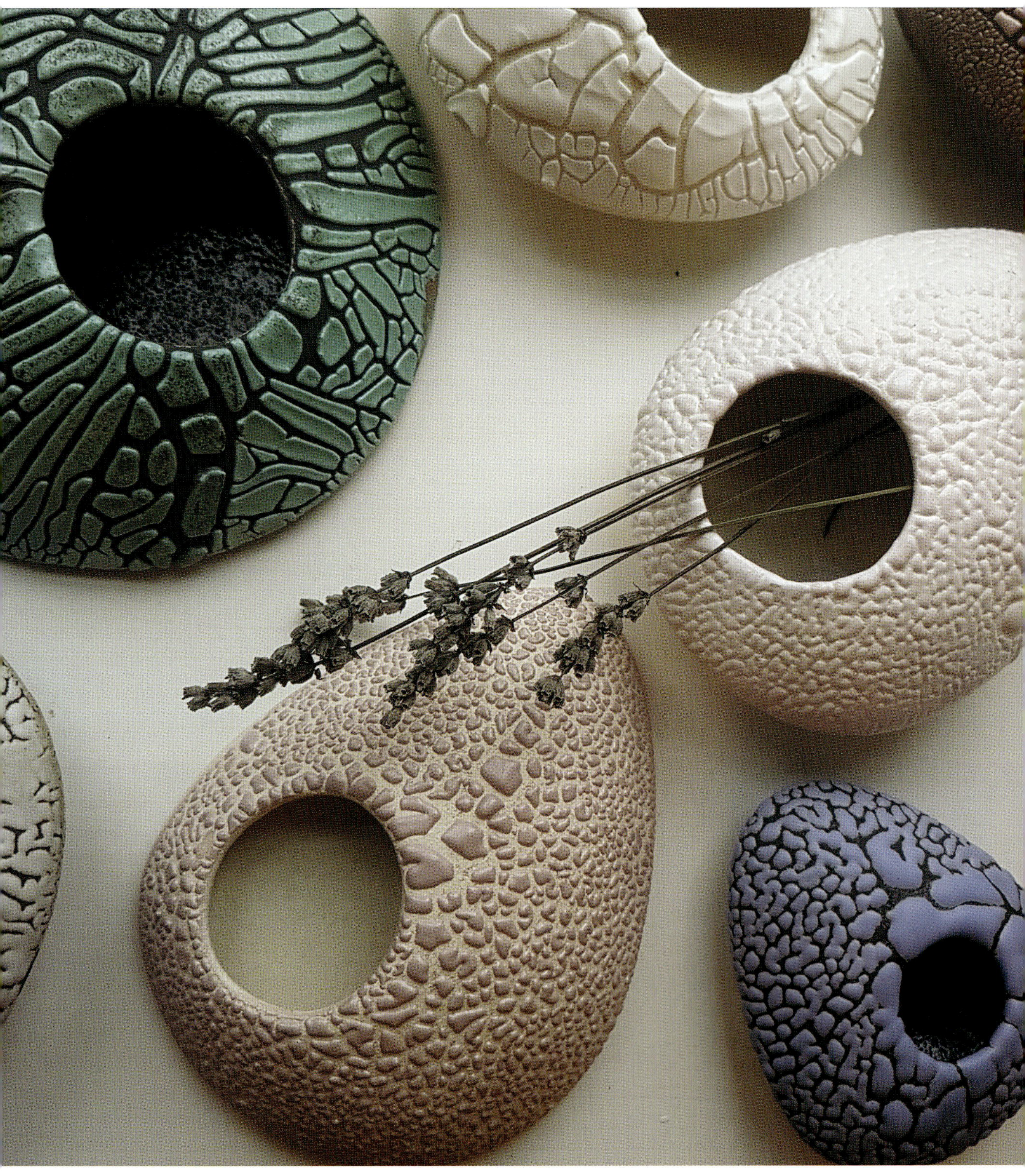

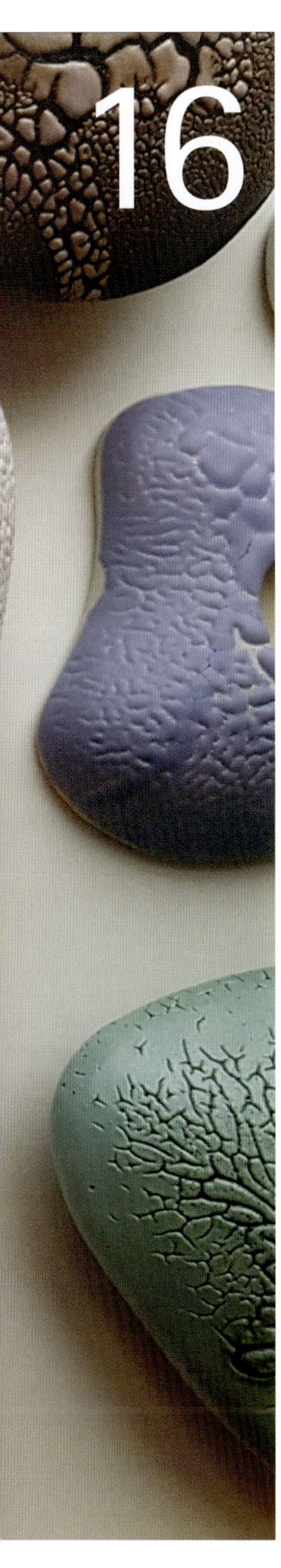

16

Shrink and Crawl: Lichen glazes

Crawling occurs when a thickly applied glaze dries to form cracks like a dried-up river bed. During firing, the cracks grow wider, leaving islands of glaze with bare clay in between. Shrink-and-crawl or lichen glazes are made by adding around 30% light magnesium carbonate, ball clay or zinc oxide to a glaze. These are light, fluffy materials that shrink on drying, forming cracks which widen during firing. These materials have a very small particle size and the particles are loosely held together, so they shrink and consolidate into islands during firing. Magnesium carbonate often forms dry, cracked lichen glazes, which can be softened by adding a small amount of zinc oxide or frit, while a combination of china clay and zinc oxide will melt to form glossy beads if fired to a high-enough temperature (usually a few cones higher than recommended).

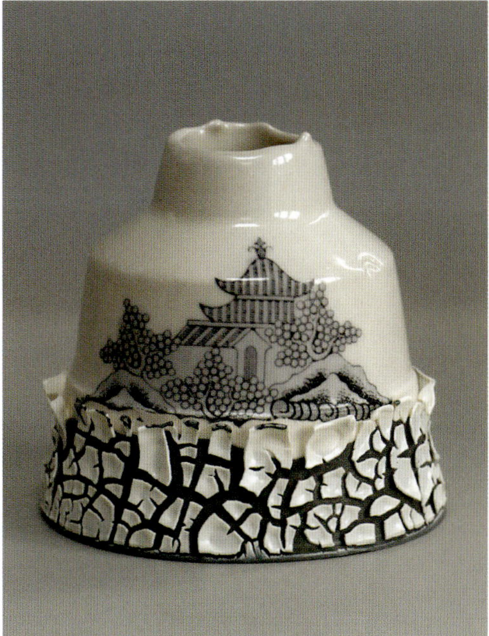

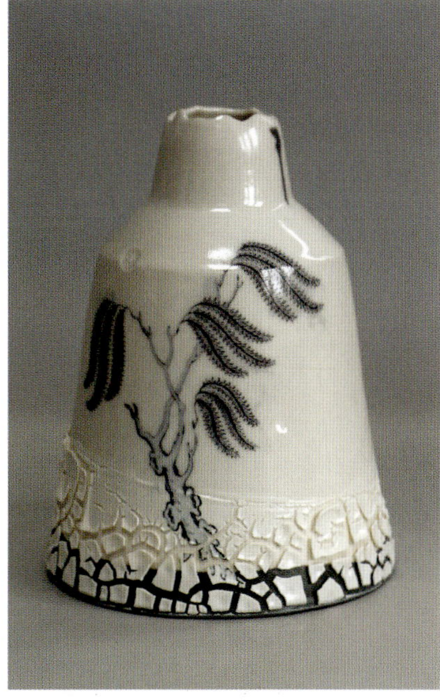

ABOVE LEFT: Raewyn Harrison, *Black Willow*, slipcast porcelain, transparent glaze, black underglaze, crawl glaze and willow pattern decal.
ABOVE RIGHT: Raewyn Harrison, *Fragments of Willow*, slipcast porcelain, transparent glaze, black underglaze, crawl glaze and willow pattern decal.

LEFT: Casja Carlenius, *Wall Vases* and vases in blue, pink, white and turquoise, height: 5cm (2in.), width: 8cm to 18cm (3¼in. to 7in.), 2018, Stockholm.

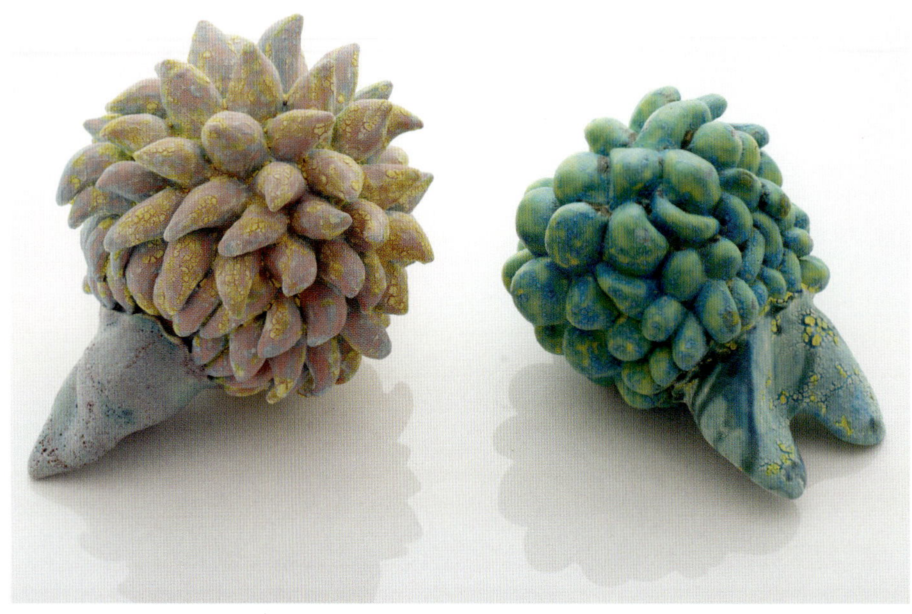

RIGHT: Linda Bloomfield, *Ice Floe Plate*, lichen glaze over transparent turquoise, fired to cone 8.

LEFT: Tessa Eastman, *Spikey & Blobby Subservient Creatures*, 2015. 22 x 20cm (8 x 7⅞in.). Private Collection. *Sylvain Deleu Photography.*

Around 30% china clay mixed with feldspar and nepheline syenite can form the basis of a simple shino, a white to orange reduction-fired glaze which often crawls. There are many types of shino, some including soluble soda ash to encourage carbon trapping. Shino glazes are traditionally fired in wood or gas kilns and require high firing temperatures.

Crawl glazes need to be applied thickly to encourage cracking to occur. The network of cracks becomes finer and closer together if the glaze is applied more thinly. Care should be taken when handling the unfired, glazed pot, as the cracked glaze can easily be knocked off when placing in the kiln. To provide contrast, these glazes can be applied on top of a dark clay body, slip, underglaze or a contrasting glaze.

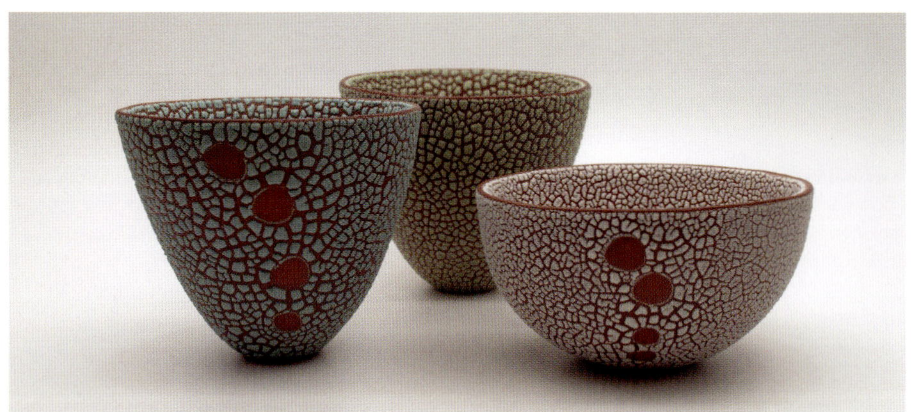

LEFT: Emma Williams, *Fingerprint Resist Bowls*, red earthenware with crawl glazes, fired to 1055°C (1931°F).

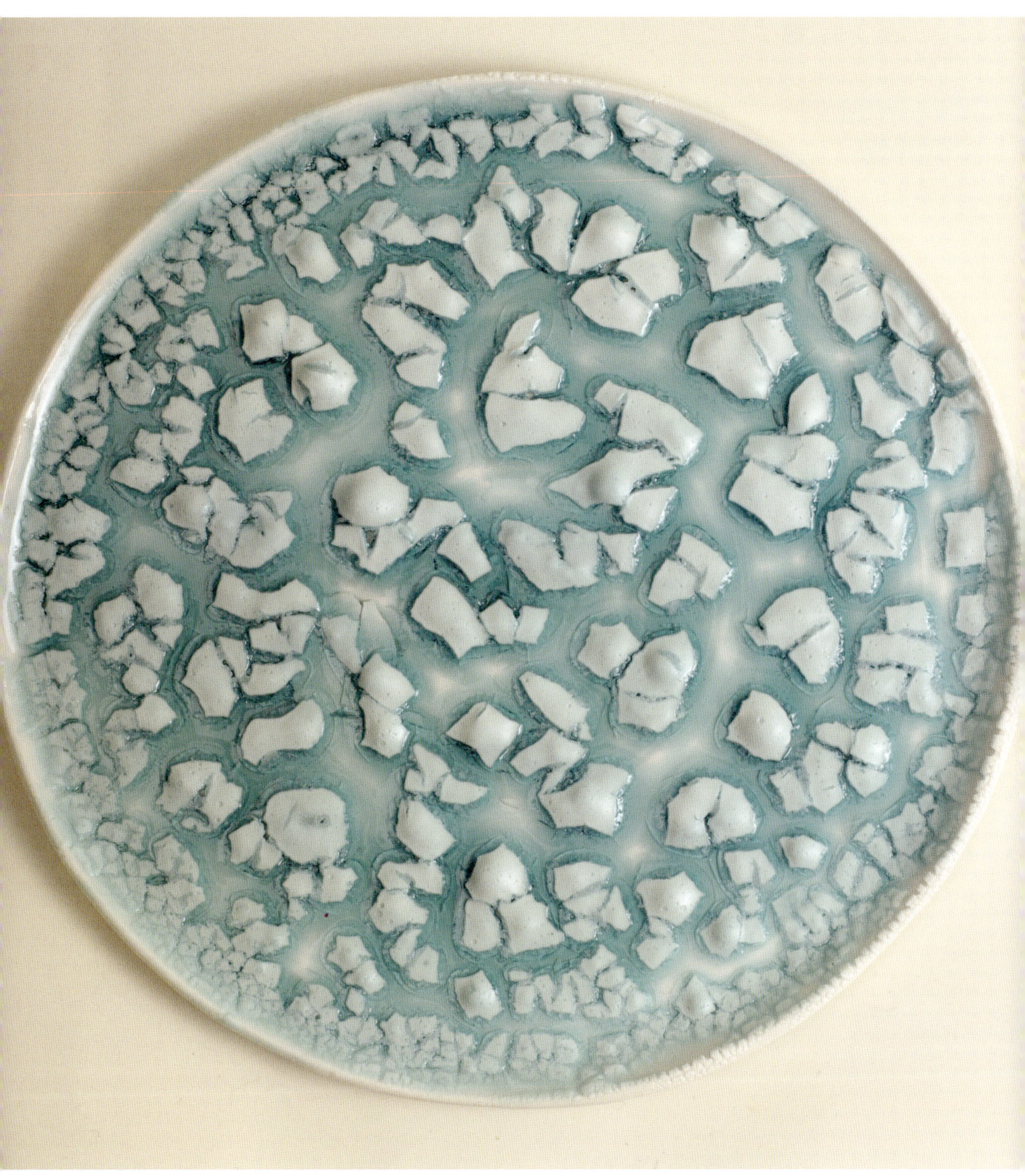

These three types of crawl glaze – magnesium, zinc and shino – are shown on the Stull map to fall along the 1:4 matt line; the further to the right along the line, the higher the firing temperature. This shows that magnesium crawl glazes can work well at low firing temperatures, zinc crawl glazes are suited to mid-range stoneware temperatures (cone 6–9), while Shino glazes need much higher firing temperatures (cone 10–12). Many Shino glaze recipes have even higher alumina – above 1.0 on the map.

RIGHT: Test tiles Set 1 showing the Lichen base glaze with the following colour additions: Top row: nickel 1, copper 1. Middle row: cobalt 0.5, iron 2. Bottom row: cobalt 0.5+ iron 2, vanadium 10.

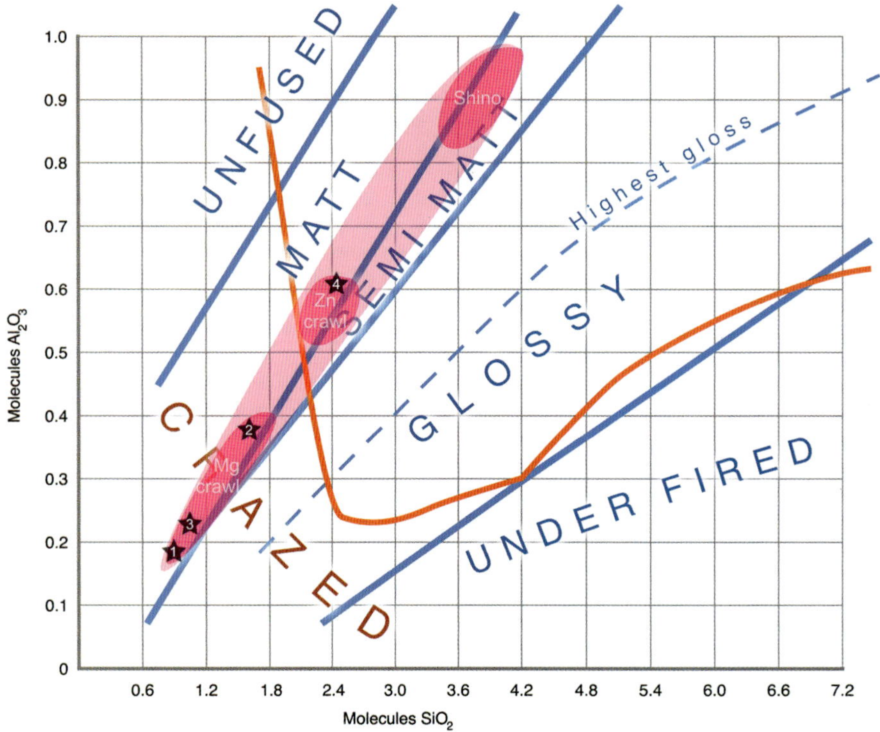

LEFT: In the Stull map, three types of crawl glazes are shown: magnesium crawl, zinc crawl and Shino. They all lie along the 1:4 matt line. The numbers refer to the glazes on pp 121–22. Graph of alumina and silica in porcelain glazes fired to cone 11 with constant flux 0.3 K2O and 0.7 CaO. The ratio of 1:5 alumina to silica gives a semi-matt glaze, while 1:8 gives a shiny glaze. The straight lines on the chart represent alumina:silica ratios of 1:4 (matt), 1:5 (semi-matt) and 1:12 (shiny, crazed glaze). The dashed line is 1:8 Al2O3:SiO2 (bright, shiny glaze). (Data from R.T. Stull 1912. Graphic by Henry Bloomfield.)

Colouring oxides can be added to the magnesium carbonate glaze type, including cobalt, copper, nickel and iron oxide. Nickel will give a pale green in magnesium crawl glazes, similar to the colour of lichen. Chromium oxide added to the zinc crawl glaze will result in a pale pink colour, but added to a magnesium crawl glaze it will turn brown.

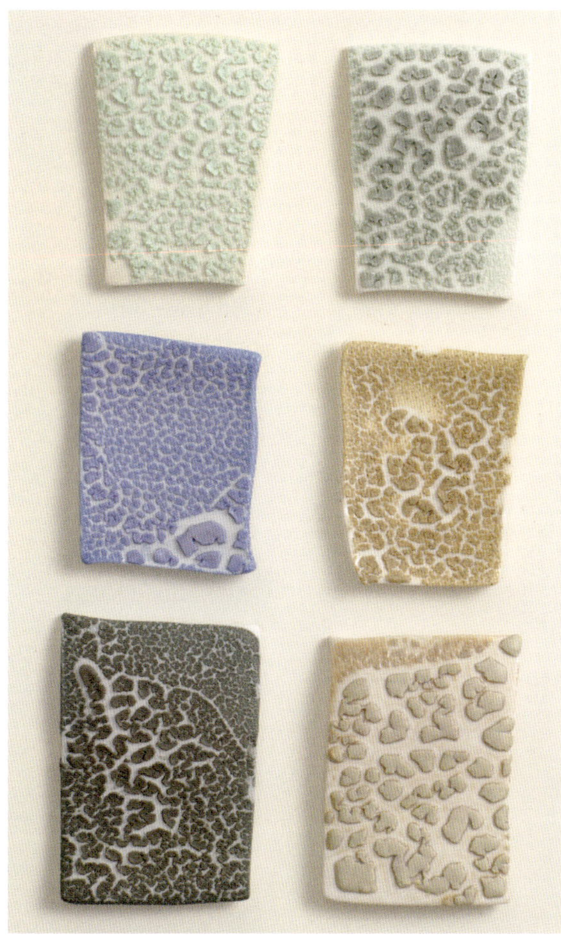

These tests show the base glaze with six colour variations.

1 Lichen glaze cone 8, 1250°C (2282°F). Glaze 1 on the Stull map.

Nepheline syenite 50
Light magnesium carbonate 40
Standard borax frit 10
+
Nickel oxide 1
Copper oxide 1
Cobalt oxide 0.5
Red iron oxide 2
Cobalt oxide 0.5 + red iron oxide 2
Vanadium pentoxide 10

Lichen glaze, cone 04, 1060°C (1940°F).

Borax frit 50
Light magnesium carbonate 30
China clay 20
+
Titanium dioxide 8

2 Lichen glaze, cone 6, 1240°C (2264°F). Glaze 2 on the Stull map.

Apply thickly
Nepheline syenite 70
Light magnesium carbonate 25
Ball clay 5
+
Zirconium silicate 5

Shino, cone 10, 1300°C (2372°F), reduction (Lisa Hammond).

Soda feldspar 70
Ball clay 30

Recipe numbers correspond to numbers marked on Stull map opposite.

3 Lichen glaze (Robin Hopper), cone 6–8, 1240–1260°C (2264–2300°F). Glaze 3 on the Stull map.

Apply thickly
Soda feldspar 30
Magnesium carbonate 31
Borax frit 6
Talc 8
Zinc oxide 6
Kaolin 19

Anderson Ranch shino slip, cone 10, 1300°C (2372°F), reduction.

Nepheline syenite 36
Kaolin 28
Spodumene 12
Ball clay 12
Soda feldspar 9
Soda ash 3
+ Bentonite 3

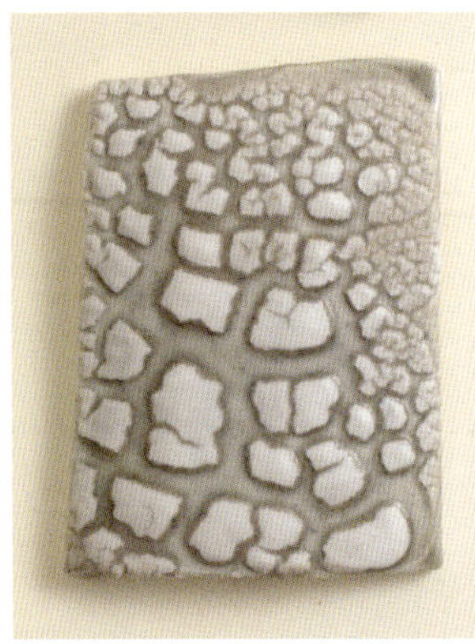

Lichen glaze 1 over satin matt grey glaze, cone 8, 1250°C (2282°F).

Satin matt grey glaze, cone 8, 1250°C (2282°F).

Potash feldspar 33
Talc 21
Whiting 12
Quartz 16
China clay 15
Zinc oxide 3
+
Cobalt oxide 2
Red iron oxide 2
Manganese dioxide 2
Nickel oxide 1

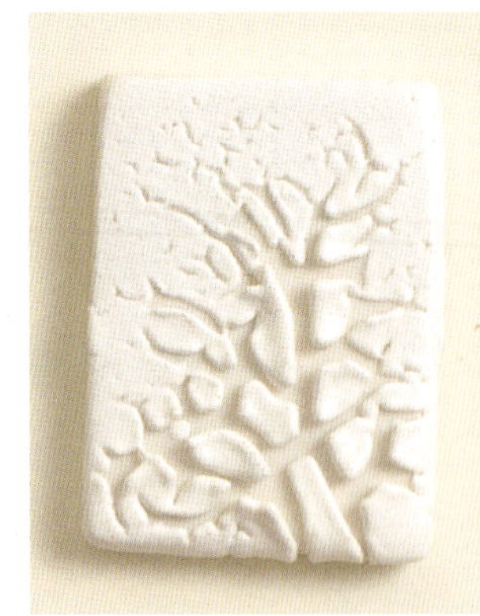

4 Zinc crawl glaze (Matt Katz), cone 8–10, 1260–1280°C (2300–2336°F). Glaze 4 on the Stull map.

(Apply thickly.)
Nepheline syenite 39
Zinc oxide 19
Flint 11
China clay 32

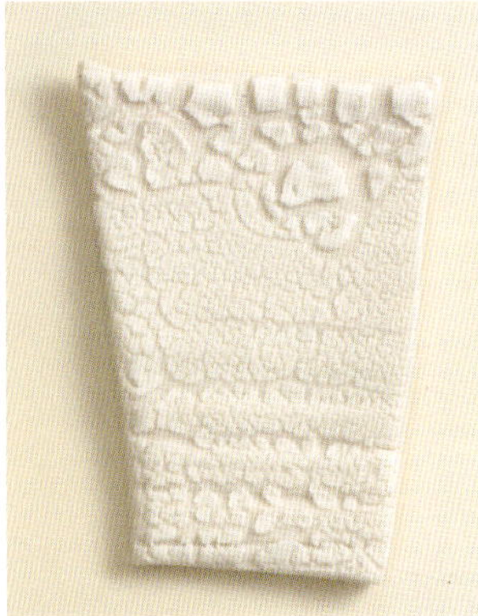

1 Lichen glaze, cone 8 1250°C (2282°F). Glaze 1 on the Stull map.

Nepheline syenite 50
Light magnesium carbonate 40
Standard borax frit 10

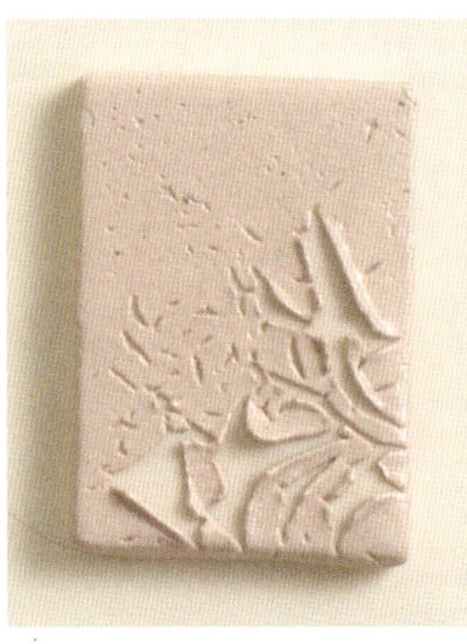

Pink crawl, cone 8–10.

Zinc crawl glaze
+ chromium oxide 0.5

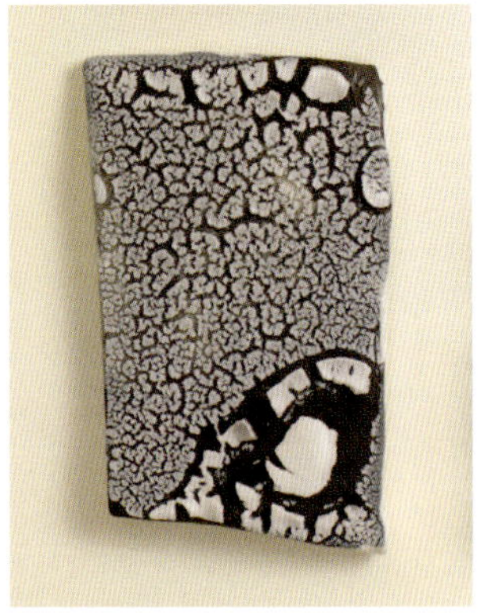

Lichen glaze 1 over black stain.

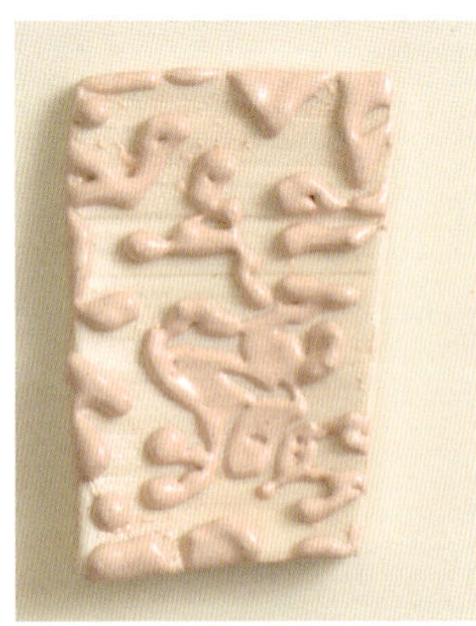

Pink beads, cone 8.

Zinc crawl glaze
+ chromium oxide 0.5
+ calcium borate frit 2

Recipe numbers correspond to numbers marked on Stull map on p.120.

Virginia Scotchie, *Chrome Knob*, 2017. Stoneware clay, textured glaze, mid-range-fired, 30 x 20 x 20cm (12 x 8 x 8in.).

Scotchie Textured Green glaze, cone 3–8, 1150–1250°C (2102–2282°F).

(Gives off fluorine gas during firing. Ventilate well.)
Bone ash 77.3
Cryolite 13.7
Soda feldspar 8.6
Barium carbonate 0.4
+ Chromium oxide 2

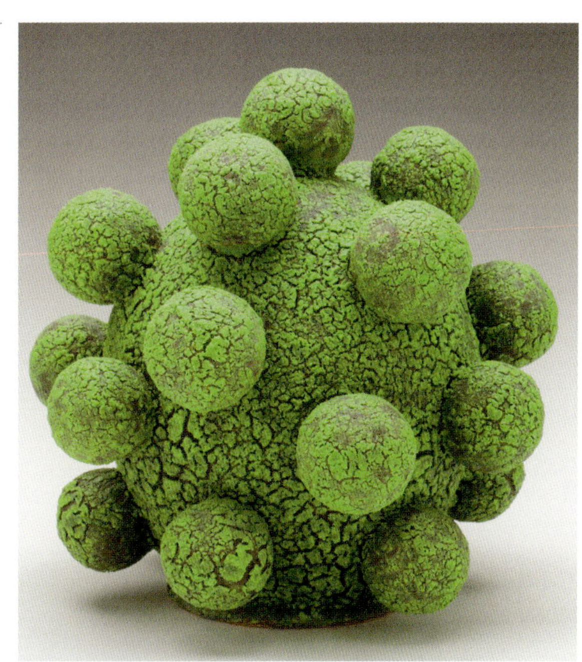

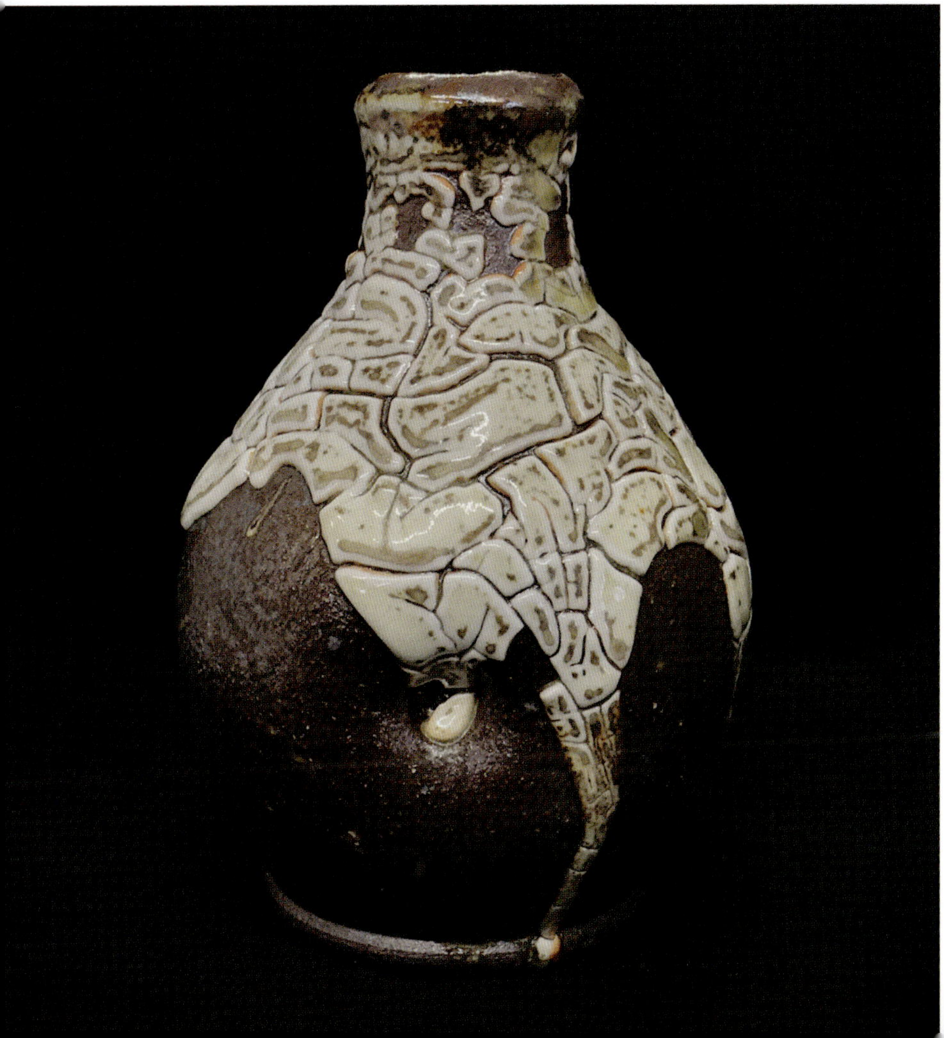

LEFT: Hollis Engley, *Dark Crawled Bottle* with Anderson Ranch shino slip. The clay body is Dark Star iron-bearing clay from STARworks Ceramics in Seagrove, NC. The glaze is a shino slip recipe from Anderson Ranch in Colorado which crawls when thick, like most Shinos (recipe on p. 121). Both fired to cone 10 (and perhaps a bit higher) over six days in the Chris Gustin anagama in South Dartmouth, MA, not far from our studio on Cape Cod.

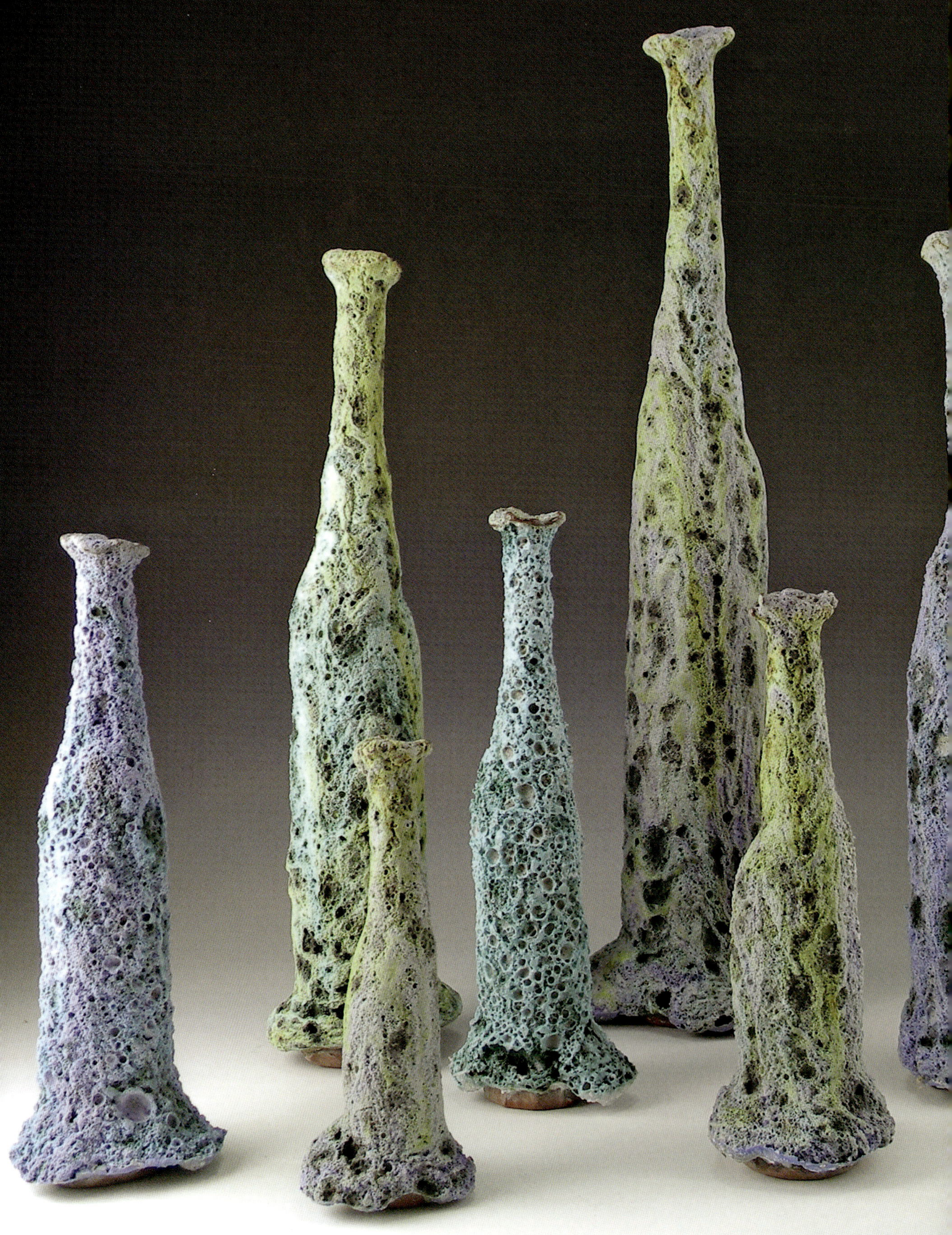

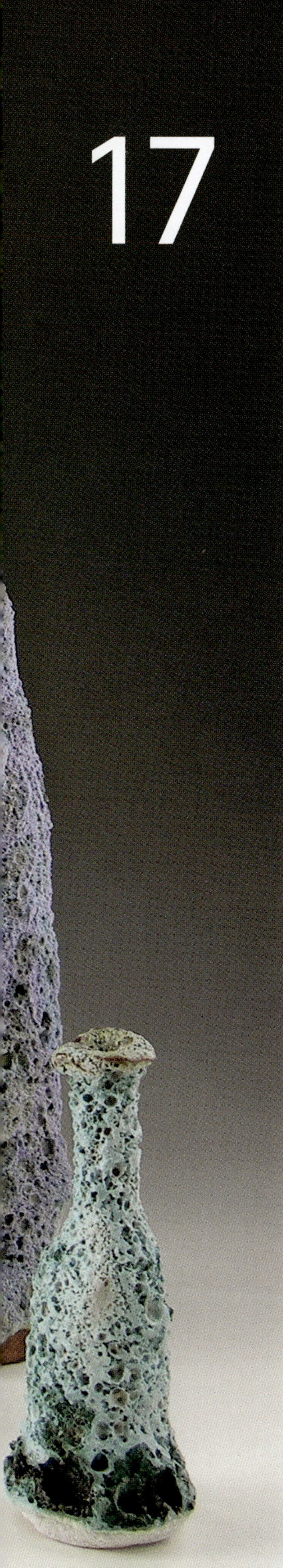

17

Volcanic, Lava or Crater glazes

Crater glazes, known as 'fat lava', were first made in Europe in the mid-20th century and by emigrant studio potters Lucie Rie in London and Gertrud and Otto Natzler in Los Angeles. Crater glazes are usually made by adding silicon carbide to a matt glaze. The silicon carbide breaks down during firing, reacting with oxides in the glaze to give off carbon dioxide. Only a small amount (0.2–2%) of silicon carbide is needed; larger quantities (2–5%) colour the glaze grey and the excess carbon dioxide gas causes the glaze to froth. Some potters use fine silicon carbide (220–1200 mesh) while others prefer coarser grades (60–120 mesh). The particle size affects the amount of gas liberated and the size and distribution of the craters. Titanium dioxide reacts readily with the silicon carbide and produces large craters if the glaze is applied very thickly. Some potters apply by brushing; the silicon carbide tends to settle quickly in the glaze bucket, so the glaze needs to be stirred frequently. The base glaze is viscous during firing so that the gas bubbles erupt and form craters without healing over. Barium and strontium matt glazes are popular as they can produce bright colours with colouring oxides (particularly copper and vanadium). Copper oxide can be reduced to copper red by the silicon carbide if there is no titanium dioxide in the glaze (compare turquoise and red glazes on p.131 and p.132). These volcanic glazes are not suitable for use on functional ware but can be used on the outside of vases and sculptural pieces.

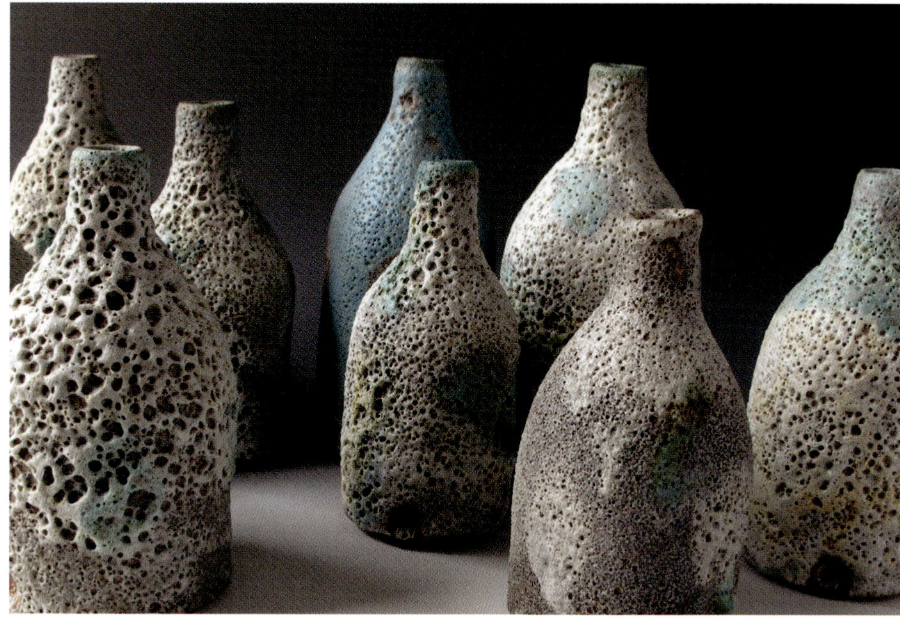

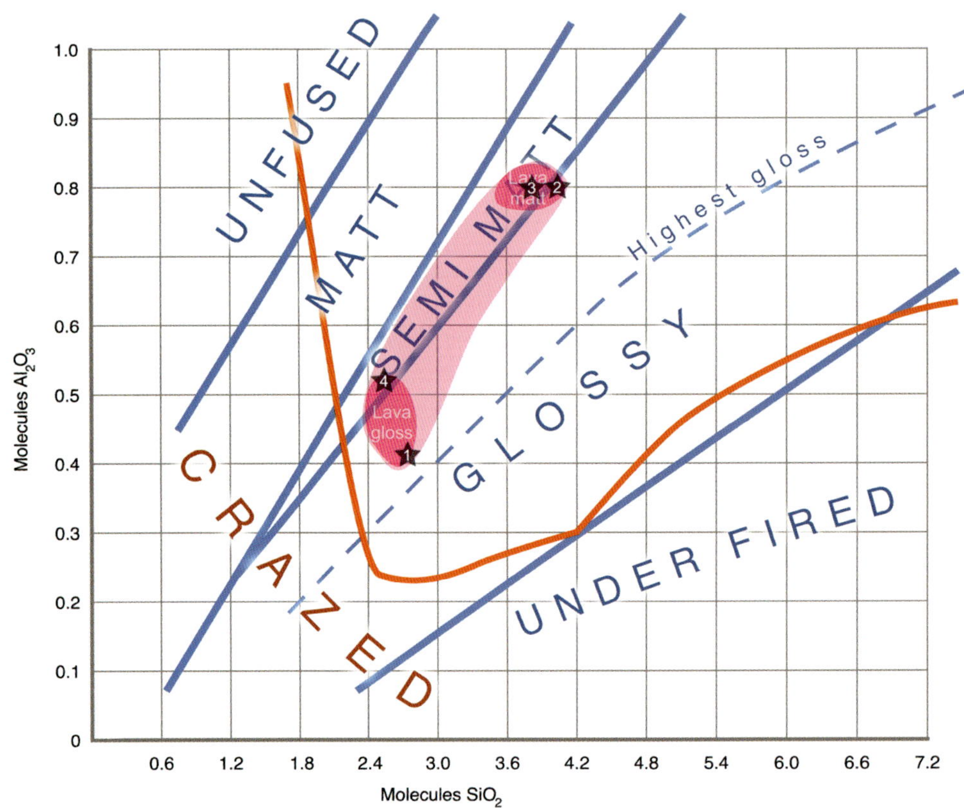

Graph of alumina and silica in porcelain glazes fired to cone 11 with constant flux 0.3 K_2O and 0.7 CaO. The ratio of 1:5 alumina to silica gives a semi-matt glaze, while 1:8 gives a shiny glaze. The straight lines on the chart represent alumina:silica ratios of 1:4 (matt), 1:5 (semi-matt) and 1:12 (shiny, crazed glaze). The dashed line is 1:8 Al2O3:SiO2 (bright, shiny glaze). (Data from R.T. Stull 1912. Graphic by Henry Bloomfield.)

Semi-glossy and semi-matt lava glazes are shown in the dark pink areas, while the light pink area shows that the range of possible lava glazes lies on or near the 1:5 semi-matt line. The numbers refer to glaze recipes on pp 127–132.

The Stull map shows that the two types of volcanic glaze are semi-glossy and semi-matt, but both lie on or near the 1:5 alumina:silica line; the further to the right along the line, the higher the firing temperature. Volcanic glazes can be layered over or under other matt glazes. A volcanic effect can also be obtained by layering a matt glaze over a clay slip or engobe containing silicon carbide. Dark-coloured clays containing iron oxide and manganese dioxide will also give off gas and react with an overlying matt glaze. Some volcanic glazes may have blisters which need to be ground down after firing. The barium matt volcanic glaze recipes given here work better at higher temperatures, around 1280°C (2336°F), as the silicon carbide needs heat and time to react with the oxides in the glaze and to break down. At lower firing temperatures, a soak at top temperature for 20–40 minutes will help liberate more gas.

Other materials that can be used as bubbling agents are cryolite, lepidolite (both of which liberate toxic fluorine gas) and vanadium pentoxide (slightly soluble and toxic). The toxicity of these materials is the reason why most potters working with volcanic glazes prefer to use silicon carbide.

PREVIOUS PAGES:

Shown on p.124, Mike Hamlin, *Bud Vase Family*, red earthenware, handbuilt from slabs and pinched. All crater glazes are brushed on using a house paint brush with 3 thick applications on the bottom ¾ of the piece and up to 10 thick applications on the top ¼ of the piece. This thick application towards the top creates the flow of the glaze. Fired in an electric kiln to 1184°C (2163°F) with slow cooling.

Shown on p.125, Katrina Pechal, bottles, thrown stoneware, biscuit slip, barium glazes, fired to cone 9.

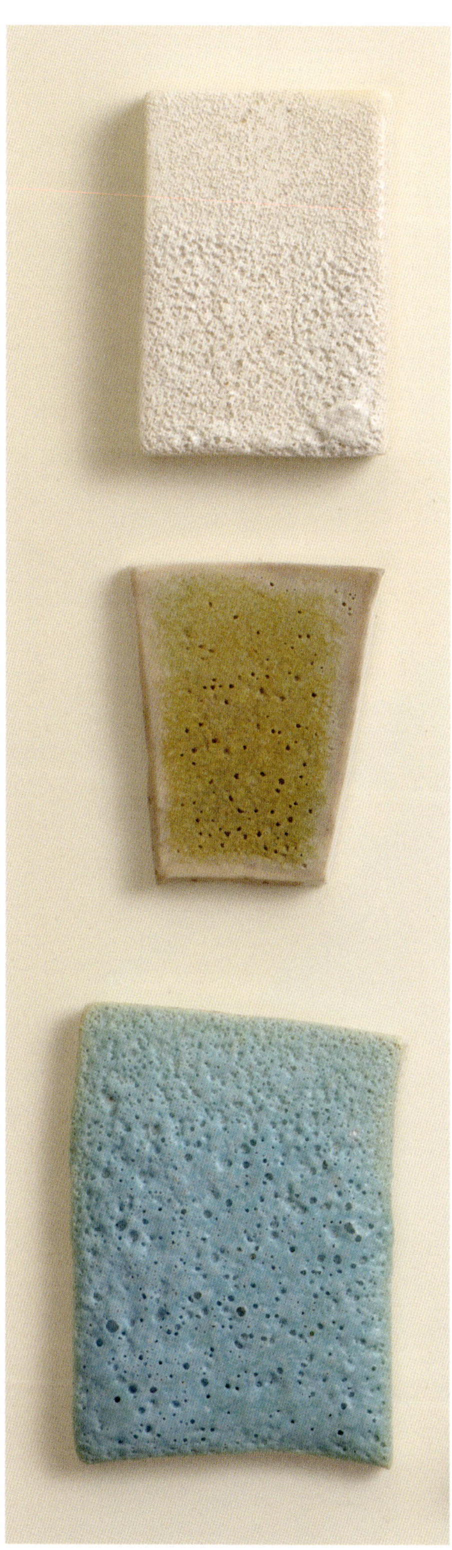

Recipe numbers correspond to numbers marked on Stull map on p.126.

1 Satin white lava glaze (from Marilee) cone 6, 1222°C (2232°F). Glaze 1 on the Stull map.

Potash feldspar 50
Whiting 24
China clay 13
Silica 13
+
Titanium oxide 11
Silicon carbide 0.3

Vanadium yellow textured matt glaze, cone 8, 1260°C (2300°F).

(Not food-safe.)

Potash feldspar 50
Dolomite 20
China clay 20
Bone ash 10
+
Vanadium pentoxide 5

4 Barium turquoise volcanic glaze cone 8, 1250°C (2282°F) over porcelain slip + silicon carbide 1%. Glaze 4 on the Stull map.
(Not food-safe.)

This glaze works better over a slip containing silicon carbide.
Nepheline syenite 55
Barium carbonate 25
Lithium carbonate 2
Flint 8
China clay 6
Calcium borate frit 5
+
Titanium dioxide 5
Copper oxide 1
Silicon carbide 1

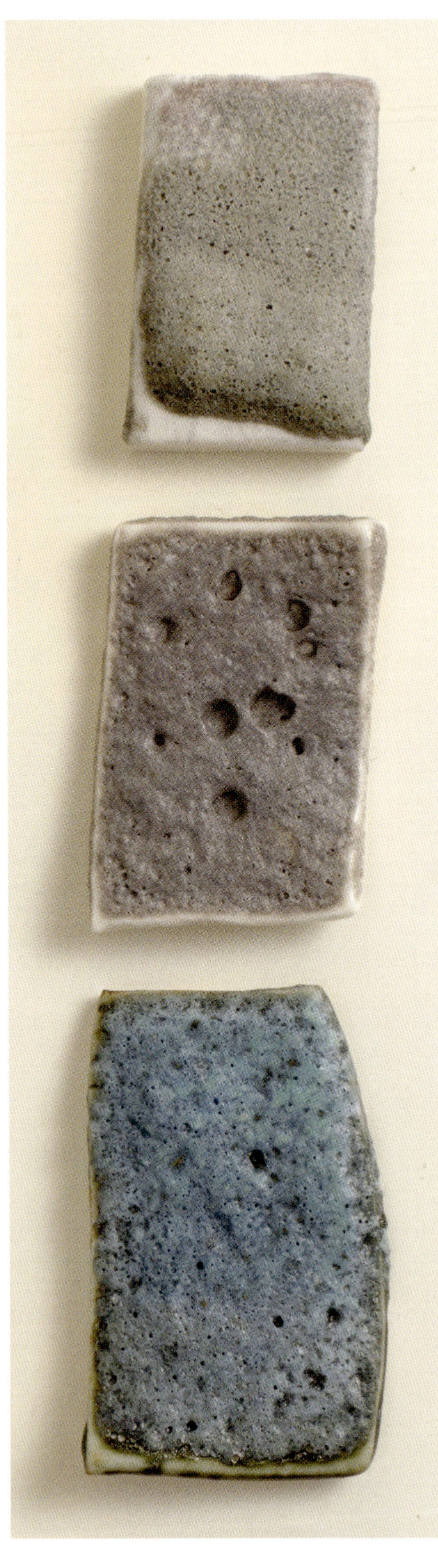

RIGHT: Satin white lava glaze tests (glaze no.1) with various additions.
Top small white tile: lava glaze fired at cone 6, 1222°C (2232°F).
Large tile on left: lava glaze fired at cone 9, 1280°C (2336°F).
Right hand column:
Satin white lava with additions:
+cobalt 0.5
+ copper 1
+chromium 0.3
+vanadium yellow 1
+vanadium yellow 8

2 Volcanic barium matt glaze (Aki Moriuchi), cone 9, 1280°C (2336°F). Glaze 2 on the Stull map.

(Not food-safe.)

Nepheline syenite 60
Barium carbonate 18
China clay 11
Quartz 13
+
Silicon carbide 4

Magnesium matt grey crater glaze, cone 8, 1250°C (2282°F).

Potash feldspar 33
Talc 21
Quartz 16
China clay 15
Whiting 12
Zinc oxide 3
+
Titanium dioxide 5
Silicon carbide 2

Barium turquoise volcanic glaze (p.127) over glossy brown-black undercoat, cone 8, 1250°C (2282°F)

Feldspar 27
Quartz 32
Whiting 21
China clay 10
+
Red iron oxide 10
Cobalt oxide 1

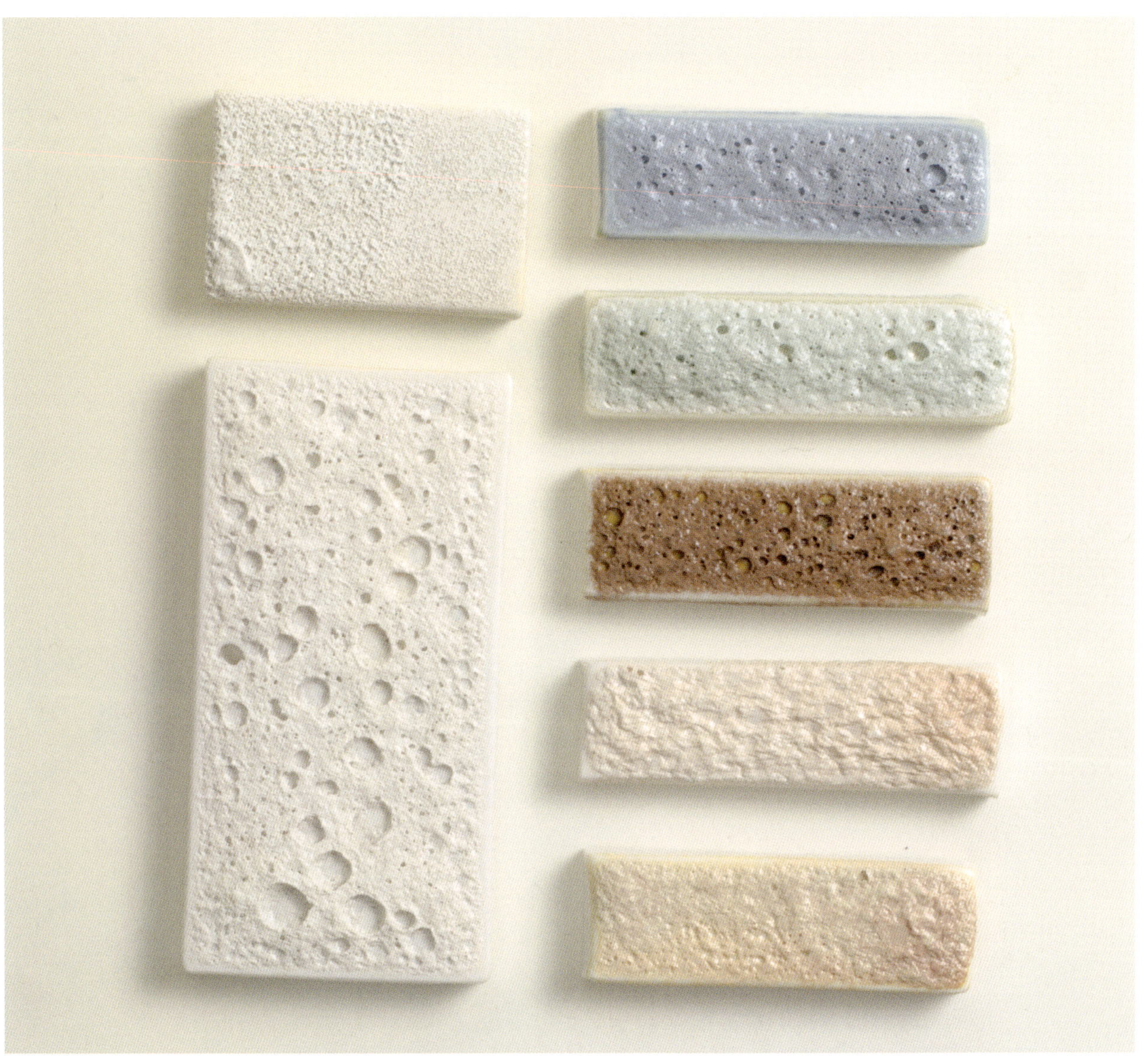

Satin white lava (Marilee's Lava), cone 9, 1280°C (2336°C), with 20-minute soak (same recipe as glaze 1 on p.127, but fired higher).

Potash feldspar 50
Whiting 24
China clay 13
Silica 13

+
Titanium oxide 11
Silicon carbide 0.3

add cobalt oxide 0.5
copper oxide 1
chromium oxide 0.3
vanadium yellow stain 1
vanadium yellow stain 8

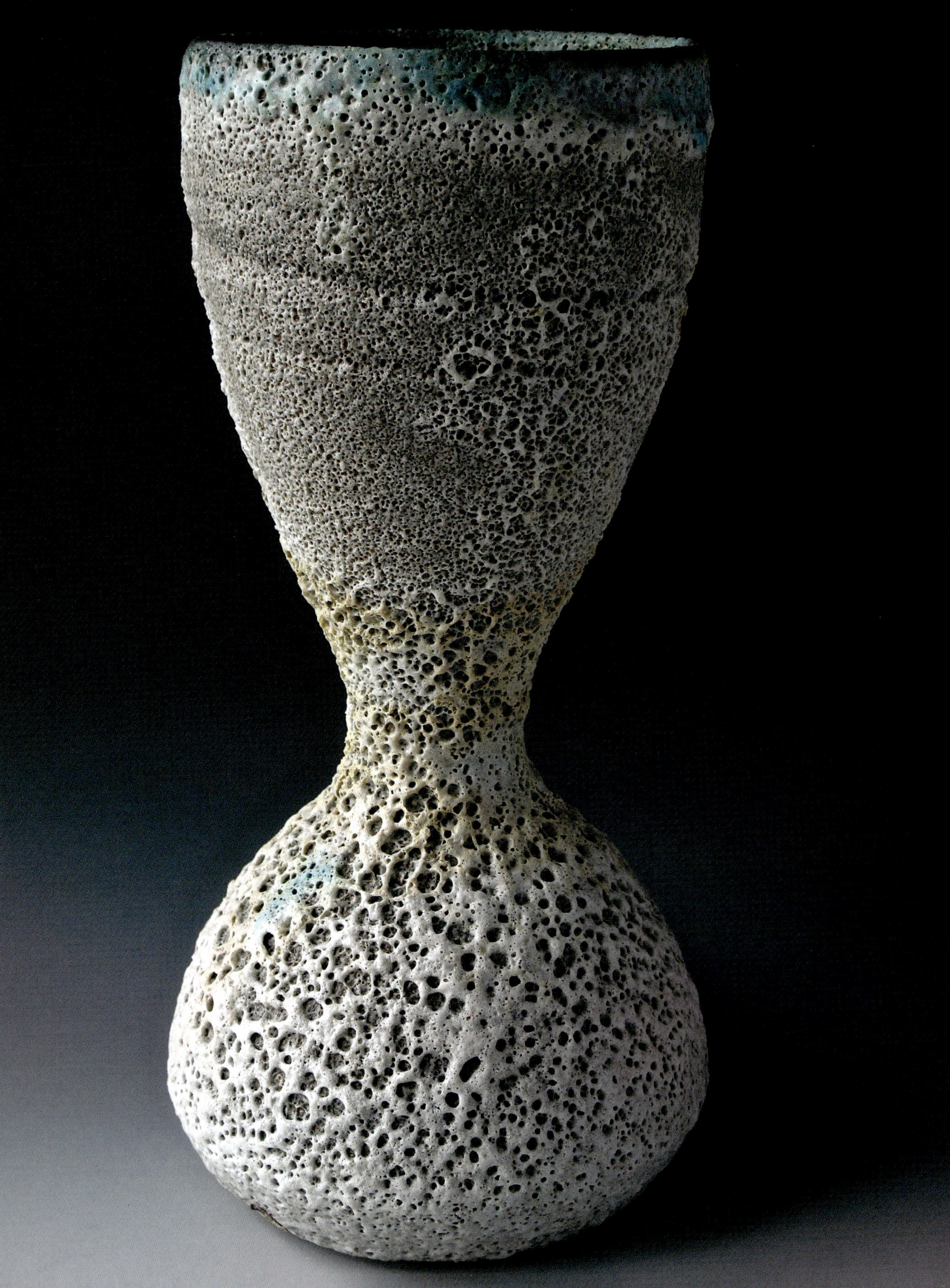

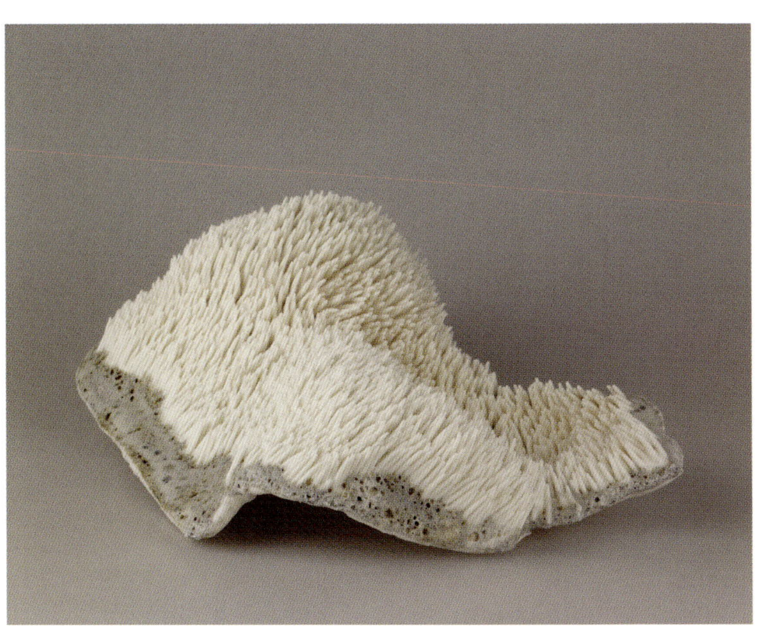

3 Barium crater glaze (Akiko Hirai), cone 9, 1280°C (2336°F) with a 20-minute soak. Glaze 3 on the Stull map.

(Not food-safe.)

Nepheline syenite 60
Barium carbonate 18
China clay 11
Quartz 10
+
Rutile 2
Silicon carbide 2

With additions of:
cobalt oxide 0.5
copper oxide 1
chromium oxide 0.3
vanadium yellow stain 1
vanadium yellow stain 8

ABOVE: Josefina Isaza, porcelain sculpture, volcanic barium matt glaze, fired to cone 9.

LEFT: Katrina Pechal, *Waisted Vase*, thrown stoneware, biscuit slip, barium glazes, fired to cone 9.

RIGHT: Volcanic glaze tests using Barium crater glaze with various additions.
Large tile: Barium Crater glaze.
Right-hand column of tiles, top to bottom:
Barium Crater Glaze
+cobalt 0.5
+copper 1
+chromium 0.3
+vanadium yelow 1
+vanadium yellow 8

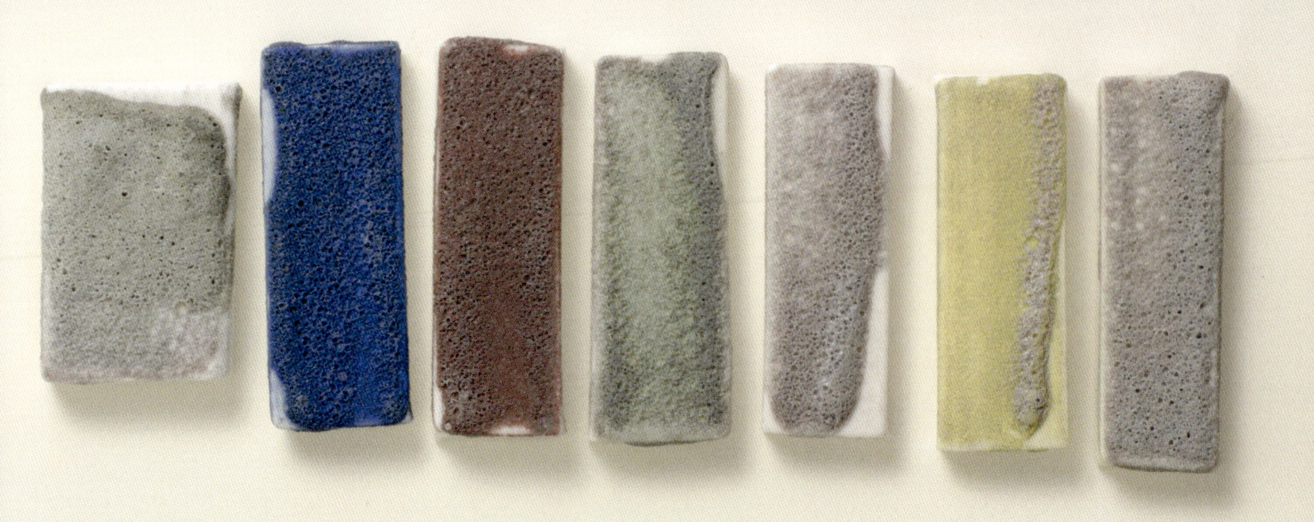

2 **Volcanic barium matt glaze (Aki Moriuchi), cone 9, 1280°C (2336°F), oxidation with 20-minute soak. Glaze 2 on the Stull map.** (Not food-safe.)

Nepheline syenite 60
Barium carbonate 18
China clay 11
Quartz 13
+
Silicon carbide 4

With additions of:
cobalt oxide 0.5
copper oxide 1
chromium oxide 0.3
vanadium yellow stain 1
vanadium yellow stain 8
praseodymium yellow stain 8

Volcanic strontium matt glaze cone 9, 1280°C (2336°F), oxidation (replacing barium with strontium).

Nepheline syenite 61
Strontium carbonate 14
China clay 11
Quartz 10
+
Titanium dioxide 2
Silicon carbide 2

ABOVE: Volcanic glaze tests using Volcanic barium matt glaze (No.2 on Stull map) with various additions. In order shown, left to right:
+cobalt 0.5
+copper 1
+chromium 0.3
+vanadium yellow 1
+vanadium yellow 8
+praseodymium yellow 8

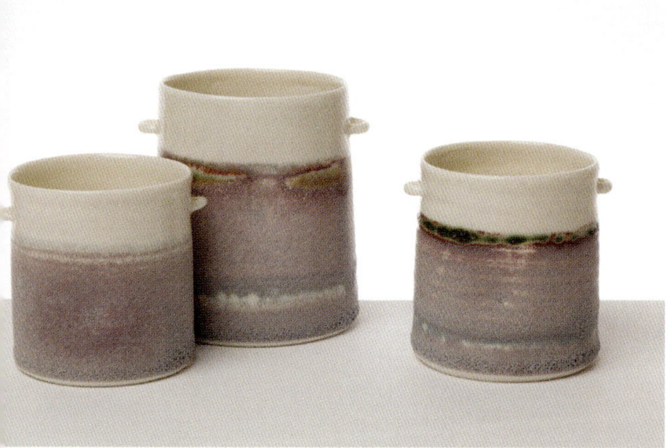

Volcanic dolomite matt glaze (Jacqui Ramrayka), cone 8, 1260°C (2300°F).

Potash feldspar 51
China clay 25
Dolomite 21
Whiting 3
+
Silicon carbide 2

LEFT: Jacqui Ramrayka, thrown porcelain with copper oxide and volcanic glaze.

RIGHT: Carys Davies, *Pebble Pot*, thrown porcelain, slip, volcanic glaze.

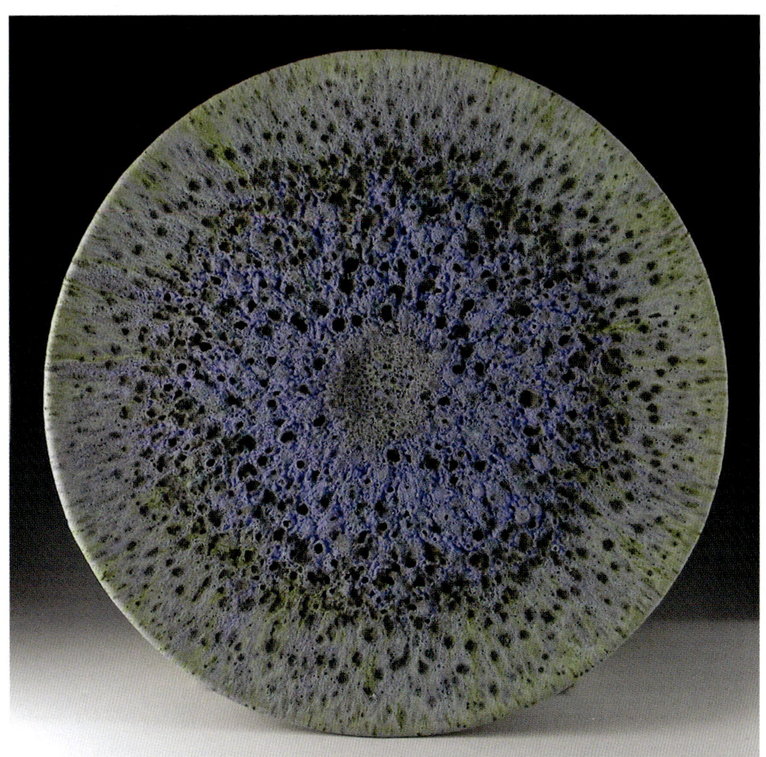

RIGHT: Mike Hamlin, *Crater Platter*, 2017. Wheel-thrown red earthenware. Crater glazes, fired in electric kiln to 1184°C (2163°F) with a slow cooling cycle, 68.5 x 5 x 5cm (27 x 2 x 2in.).

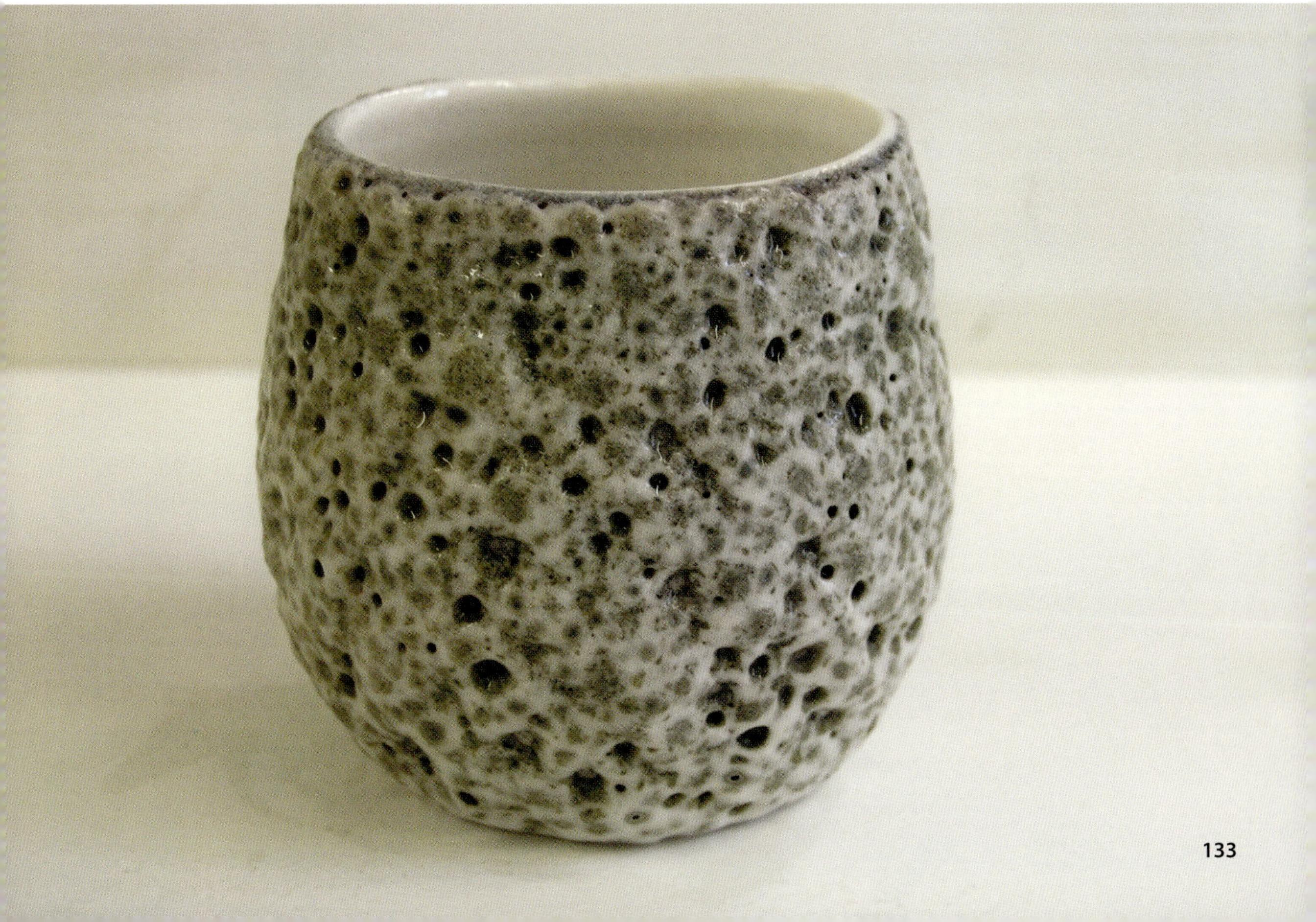

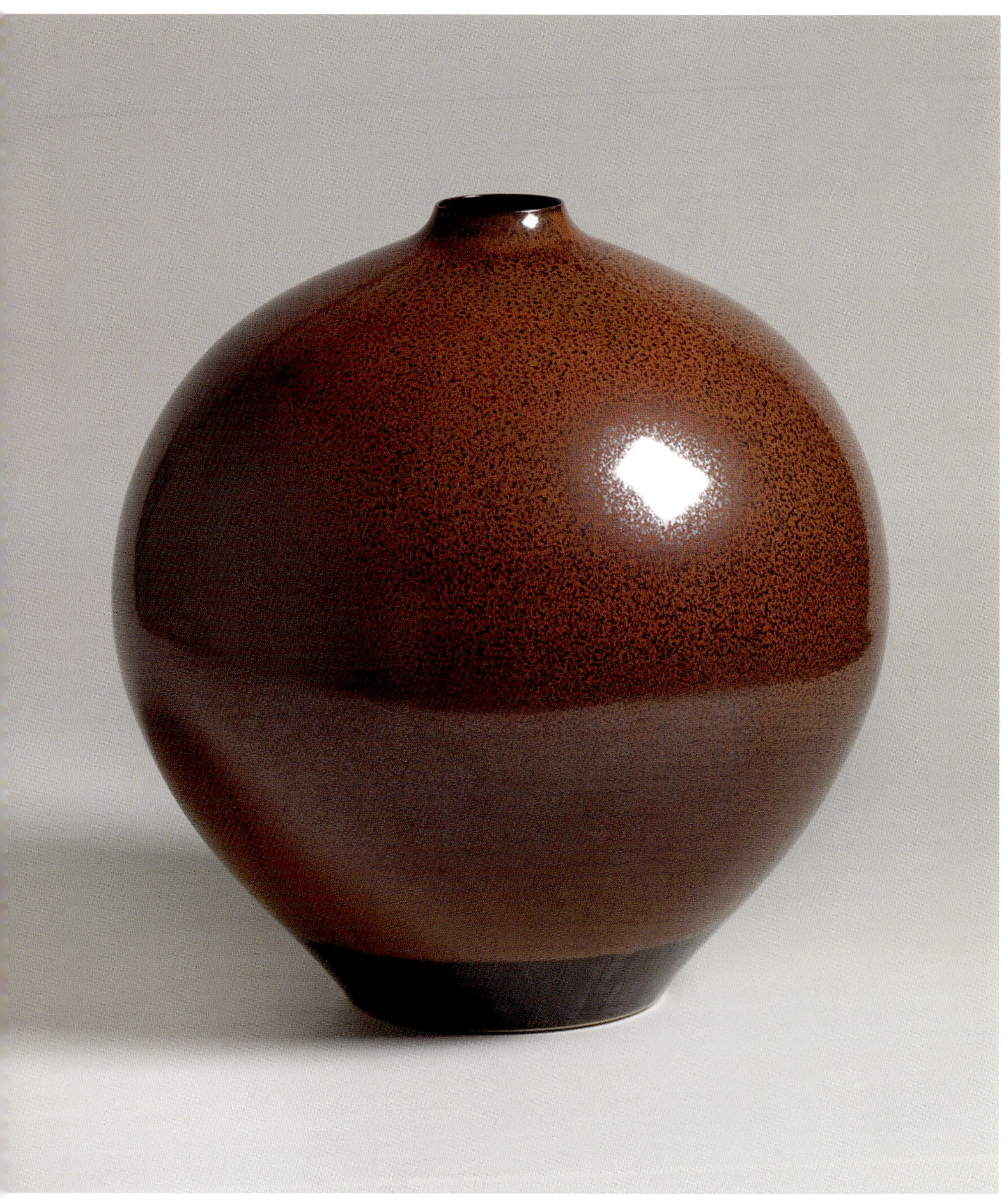

18

Spotted glazes

Oil-spot glazes are traditionally dark brown or black glazes high in iron oxide, usually containing around 6–10% red iron oxide. The red iron oxide or haematite, Fe_2O_3, changes from a trigonal crystal structure to cubic magnetite, Fe_3O_4, at 1210–1232°C/2210–2250°F (cone 6–7), releasing bubbles of oxygen gas which drag to the surface iron spots of darker magnetite. Oil-spot glazes are applied thickly and fired in an oxidation atmosphere, traditionally at cone 9–10, but they are also possible at cone 6–8. Adding magnesium in the form of talc or dolomite increases the viscosity of the molten glaze, allowing spots to form. The effect can be enhanced by applying a layer of opaque white glaze on top of the dark iron glaze. Copper oxide also releases oxygen gas at similar temperatures and can be used to make a turquoise oil-spot effect if coated with an opaque matt glaze. These types of spotted glazes are more fluid in the melt than volcanic glazes and the eruptions heal over to produce a smooth, glossy, spotted surface. They are therefore suitable for use on functional ware.

LEFT: Suleyman Saba, large iron-spot vase, stoneware with saturated iron glaze.

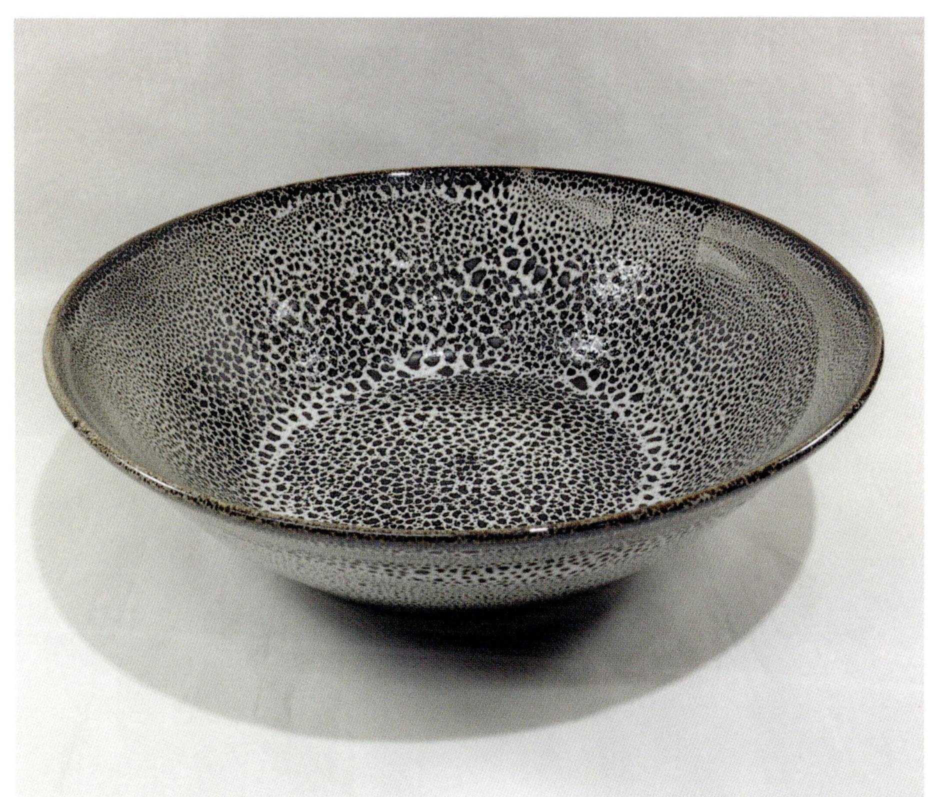

RIGHT: Annie Jennings, large stoneware bowl, fired to 1280°C (2336°F) with a 20-minute soak in an electric kiln. Using Michael Bailey's brown oil-spot glaze covered with white crackle overglaze.
Michael Bailey's brown oil-spot glaze, cone 10, 1280°C (2336°F), 20-minute soak.
Potash feldspar 26
Soda feldspar 36
Dolomite 5
Talc 5
Frit 3110 5
Quartz 8
China clay 15
+
Red iron oxide 6

White crackle overglaze cone 10, 1280°C.
Soda feldspar 83
Whiting 8
Quartz 8
+
Zirconium silicate 10

Iron red glaze (Michael Bailey), cone 6–8, 1240–1260°C (2264–2300°F), under white opaque glaze.

Potash feldspar 47
Bone ash 15
Lithium carbonate 4
Talc 17
Quartz 11.5
China clay 6
+
Red iron oxide 11.5

Glossy brown undercoat glaze, cone 8, 1250°C (2282°F), under white opaque glaze.

(This glaze is slightly runny on vertical surfaces.)
Potash feldspar 34
Quartz 23
Borax frit 14
China clay 13
Whiting 11
Dolomite 5
+
Red iron oxide 10

Runny turquoise undercoat glaze, cone 8, 1250°C (2282°F), under white opaque glaze.

(This glaze is very runny and works best on flat surfaces. It is less runny if standard borax frit is used instead of calcium borate frit.)
Potash feldspar 45
Quartz 17
Calcium borate frit 15
Whiting 14
China clay 5
+
Copper oxide 1

Glossy brown-black undercoat glaze, cone 8, 1250°C (2282°), under white opaque glaze.

(This is a stiff glaze that will not run.)
Feldspar 27
Quartz 32
Whiting 21
China clay 10
+
Red iron oxide 10
Cobalt oxide 1

Glossy black undercoat glaze, cone 8, 1250°C (2282°F), under white opaque glaze.

(Apply white glaze thinly for spotted effect, more thickly for shrink and crawl effect.)
Potash feldspar 34
Quartz 28
Borax frit 14
China clay 13
Whiting 11
Dolomite 5
+
Red iron oxide 10
Cobalt oxide 1

White opaque satin matt top coat glaze over glossy black undercoat glaze, cone 8, 1250°C (2282°F).

(The black glaze is slightly runny on vertical surface; see test tile on opposite page, bottom right.)
Potash feldspar 33
Talc 21
Whiting 12
Quartz 16
China clay 15
Zinc oxide 3
+
Zirconium silicate 5

RIGHT: Oil-spot glaze tests. Top row: Iron red+white opaque, glossy brown–black+white opaque. Middle row: Glossy brown+white opaque, glossy black+white opaque. Bottom row: Runny turquoise+white opaque, glossy black+white opaque on vertical tile. The black glossy undercoat and white opaque glaze were fired vertically on stoneware. All other tests were fired horizontally on porcelain.

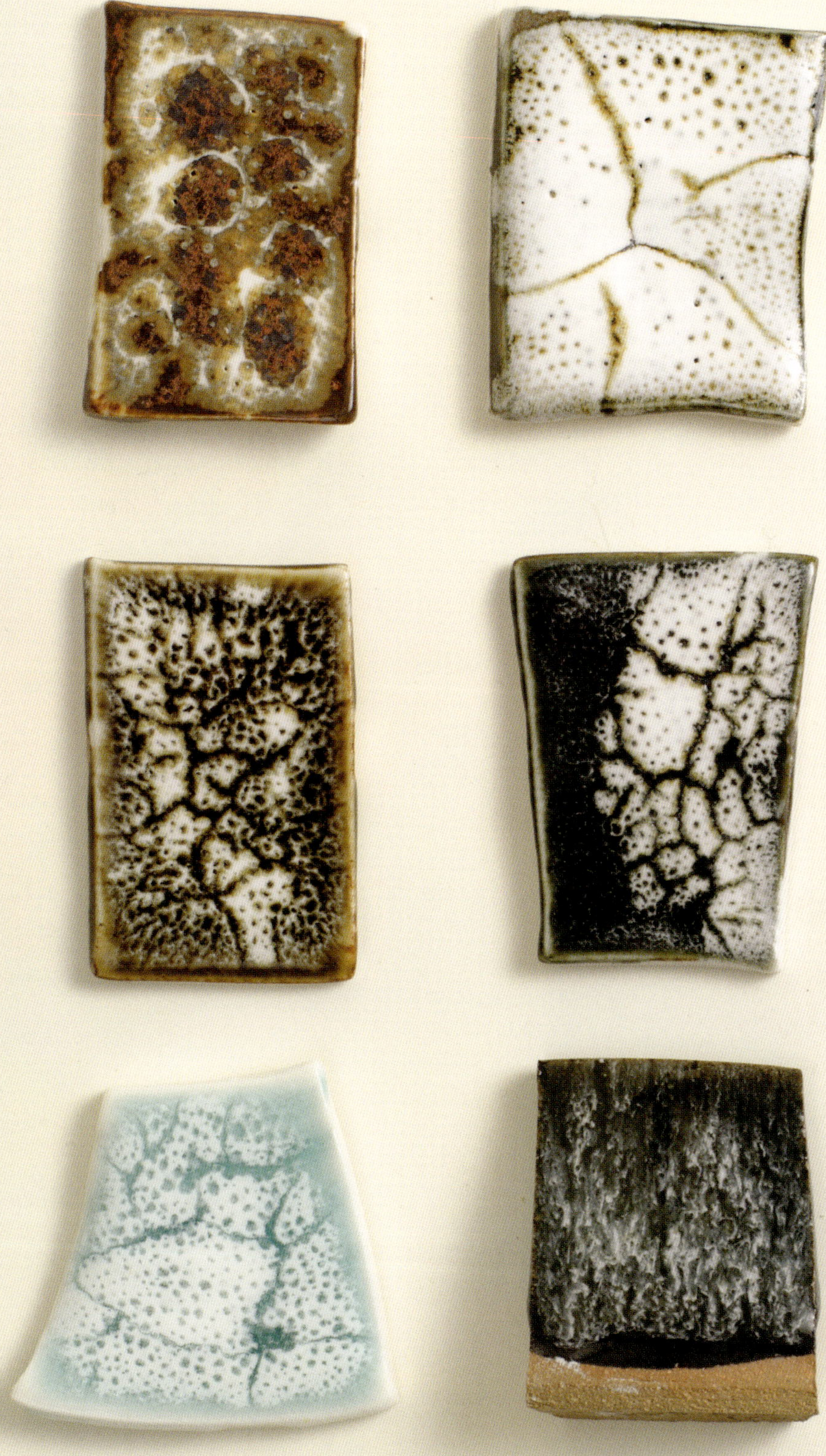

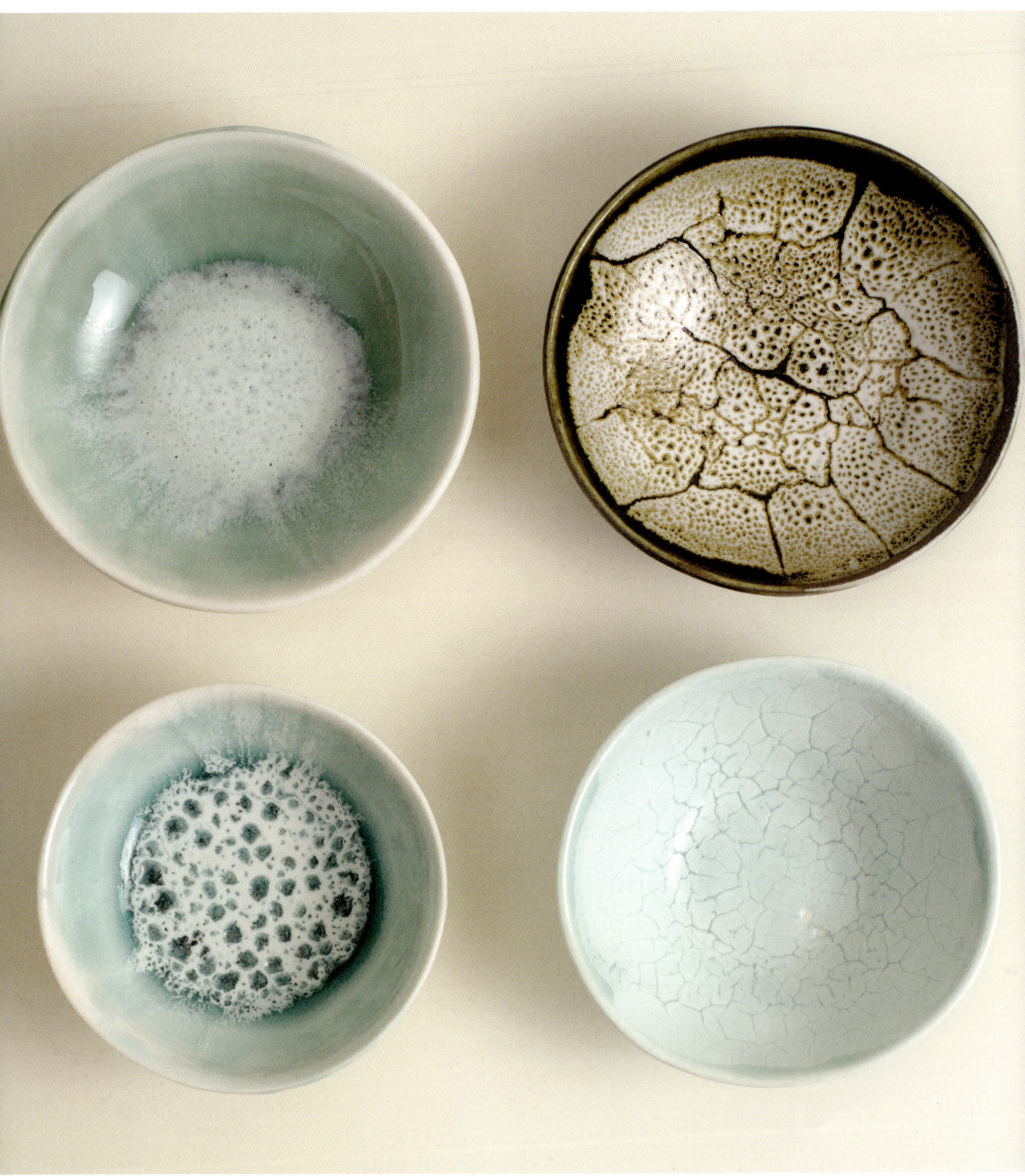

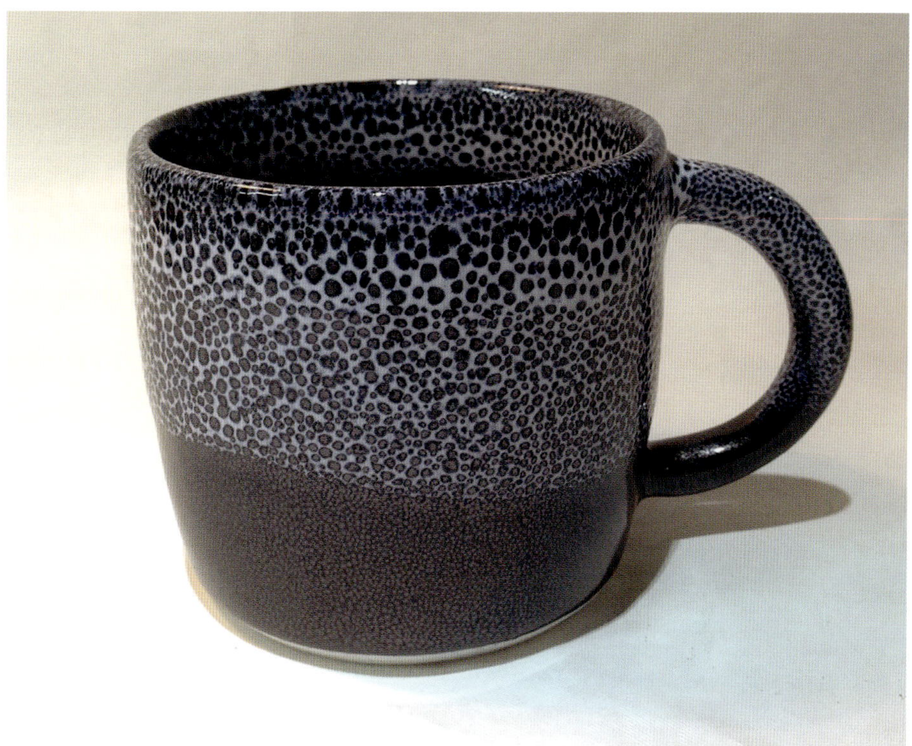

LEFT: Linda Bloomfield spotted glaze test bowls. Top left: Runny turquoise with no quartz. Top right: Glossy brown–black glaze under white opaque glaze. Bottom left: runny turquoise under white opaque glaze. Bottom right: Turquoise with added quartz 6 and china clay 8 under white talc glaze. Thrown porcelain fired to cone 8, 1250°C (2282°F).

ABOVE RIGHT: Annie Jennings, stoneware mug, fired to 1280°C (2336°F) with a 20-minute soak in an electric kiln. Michael Bailey's oil-spot glaze overlapped with white crackle overglaze.

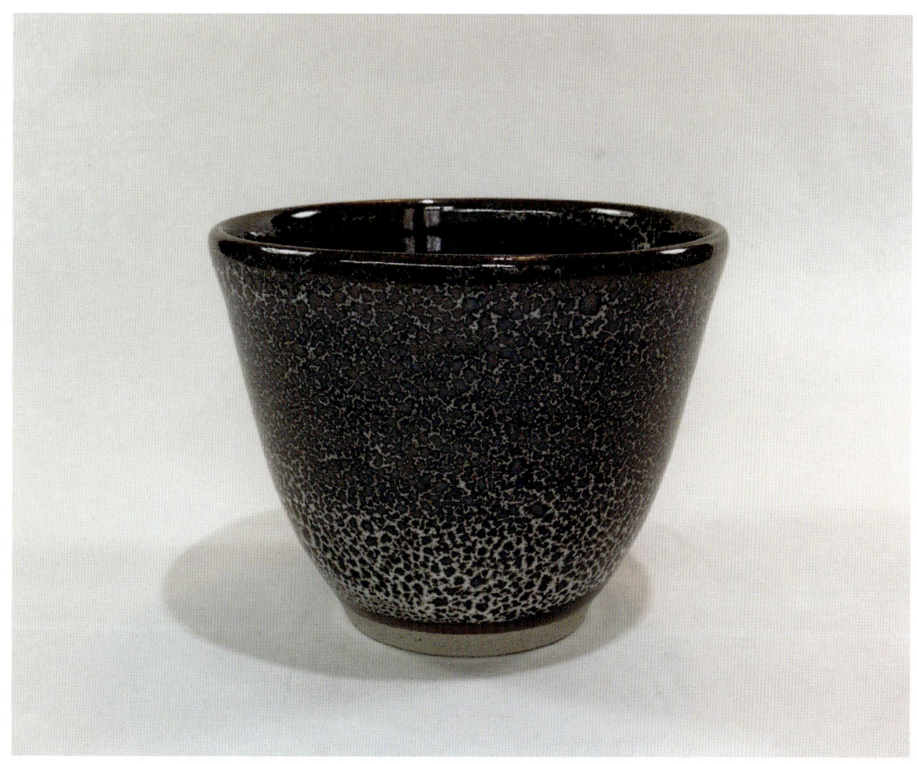

RIGHT: Annie Jennings, stoneware bowl, fired to 1280°C (2336°F) with a 20-minute soak in an electric kiln. Using Michael Bailey's brown oil-spot glaze covered with white crackle overglaze.

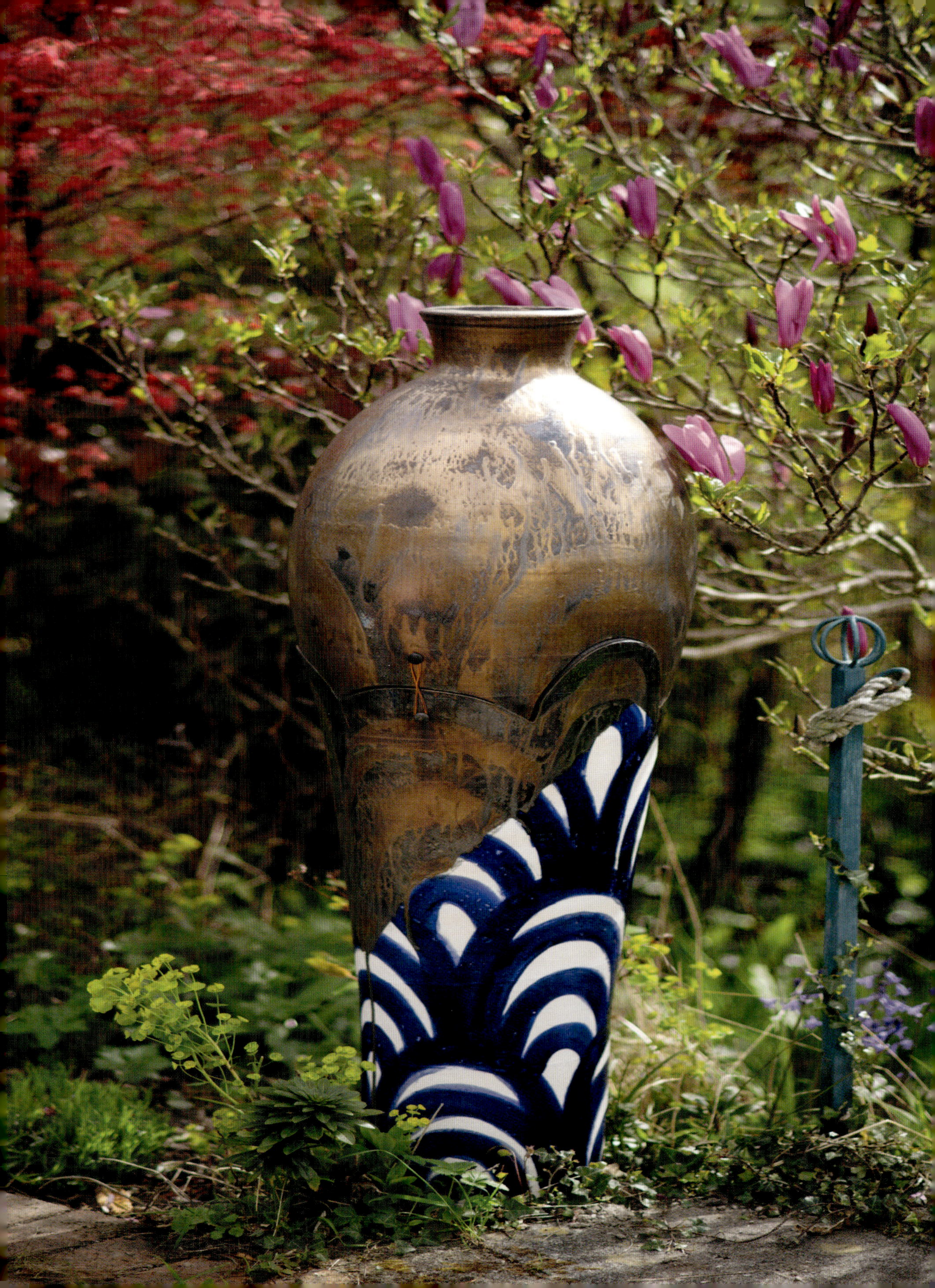

19

Metallic glazes

Some metallic glazes are classified as pigments rather than glazes, as they contain very little silica and are made from colouring oxides mixed with clay. Metallic glazes are made by saturating the glaze with colouring oxides, usually copper and manganese oxides. These glazes fire to a black colour when applied thinly and must be applied thickly to get bronze. However, they sometimes cause a wrinkled effect when too thick. The excess copper and manganese remain undissolved and form a metallic bronze effect on the surface of the glaze. Some metallic glaze recipes also use cobalt oxide, but this is expensive and it is possible to get the same bronze effect using nickel or just manganese and copper. Avoid kiln fumes from this type of glaze as they contain manganese dioxide, which is toxic. Make sure the kiln room is well ventilated. These metallic glazes are not food-safe as they are basically a manganese and copper oxide wash.

Other metallic glazes are overloaded with a mixture of colouring oxides, including manganese, iron, copper and cobalt.

RIGHT: Metallic glaze test tiles. Left: Dry metallic, right: Bronze glaze.

LEFT: Taz Pollard, Merneith garden sculpture. Ceramic with metal rivets and copper wire. The bronze glaze is a Stephen Murfitt recipe, 1.3m x 40cm (40½ x 15¾in.). *John Russel Photography.*

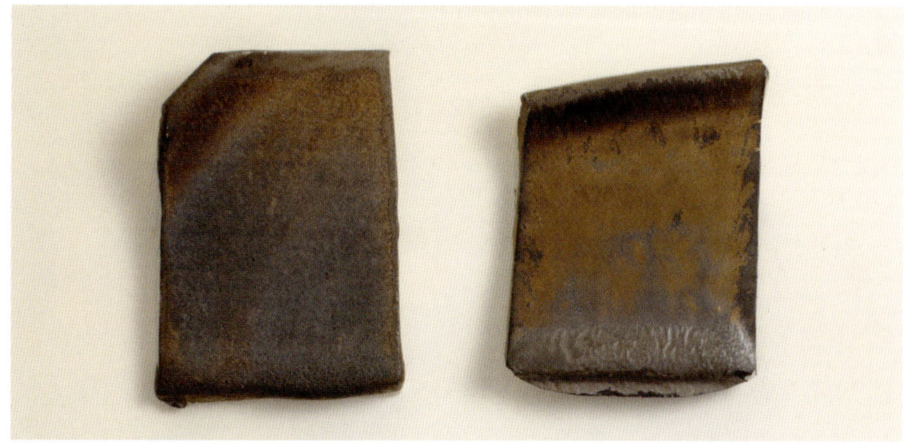

Dry metallic black/bronze glaze, cone 8–10, 1260–1280°C (2300–2336°F).

(For a metallic crystalline glaze, try increasing the feldspar to 65 and reducing the clay to 5.)

Red clay 40
Potash feldspar 30
Manganese dioxide 26
Copper oxide 4

Bronze glaze (from Steve Ogden) cone 2–6, 1160–1220°C (2120–2228°F).

Red earthenware clay 20
China clay 10
Manganese dioxide 60
Copper oxide 10

Pewter glaze (Stephen Murfitt), cone 6–8, 1240–1260°C (2264–2300°F).

Manganese dioxide 77
China clay 23,

Bronze glaze (Peter Wills), cone 8–10, 1285°C (2345°F).

FFF feldspar 24
Barium carbonate 24
Manganese dioxide 36
Copper oxide 12
Bentonite 5

Bronze glaze (Stephen Murfitt), cone 6–8, 1240–1260°C (2264–2300°F).

Manganese dioxide 61
China clay 23
Copper oxide 8
Cobalt oxide 8

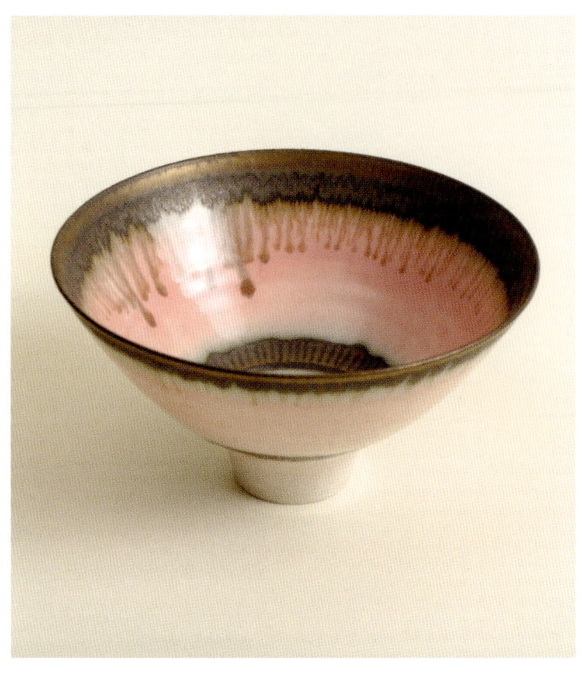

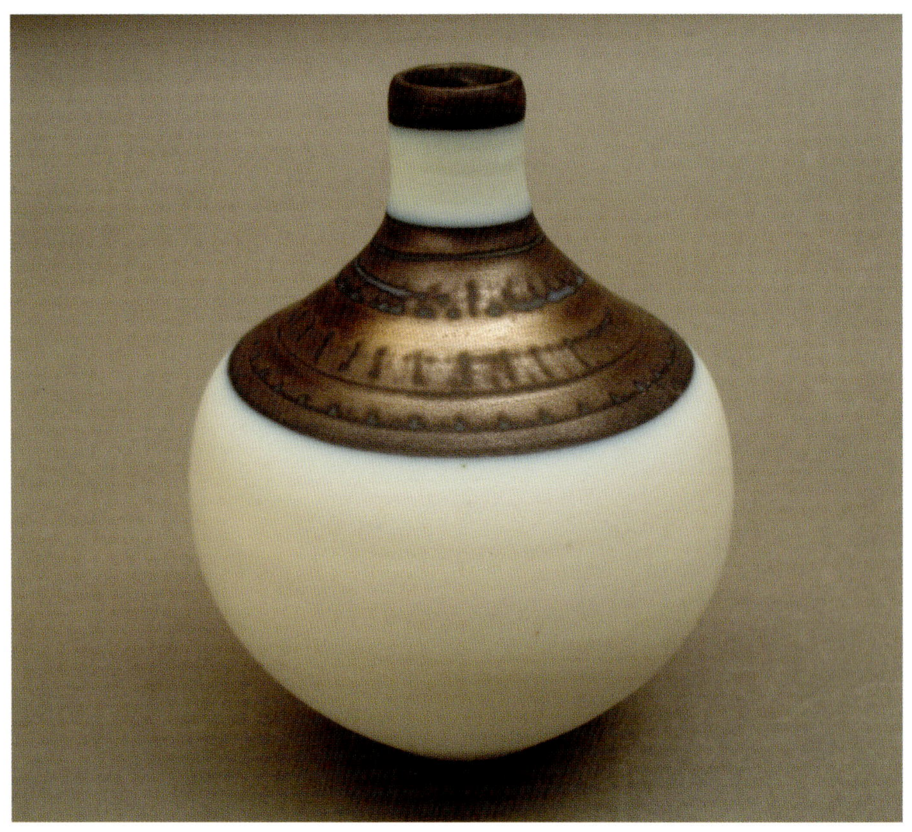

ABOVE: Peter Wills, pink and bronze porcelain bowl, raw-glazed, thrown porcelain, fired in an electric kiln to 1280°C (2336°F).

LEFT: Richard Baxter, thrown porcelain bottle, bronze glaze, fired in an electric kiln to 1240°C (2264°F).

RIGHT: Linda Bloomfield, test porcelain bowls. Top left: silicon carbide copper red glaze. Top right and bottom left: Runny pink and dry metallic bronze glaze. Bottom right: Satin matt talc white and dry metallic bronze glaze. Fired to cone 8, 1250°C (2282°F).

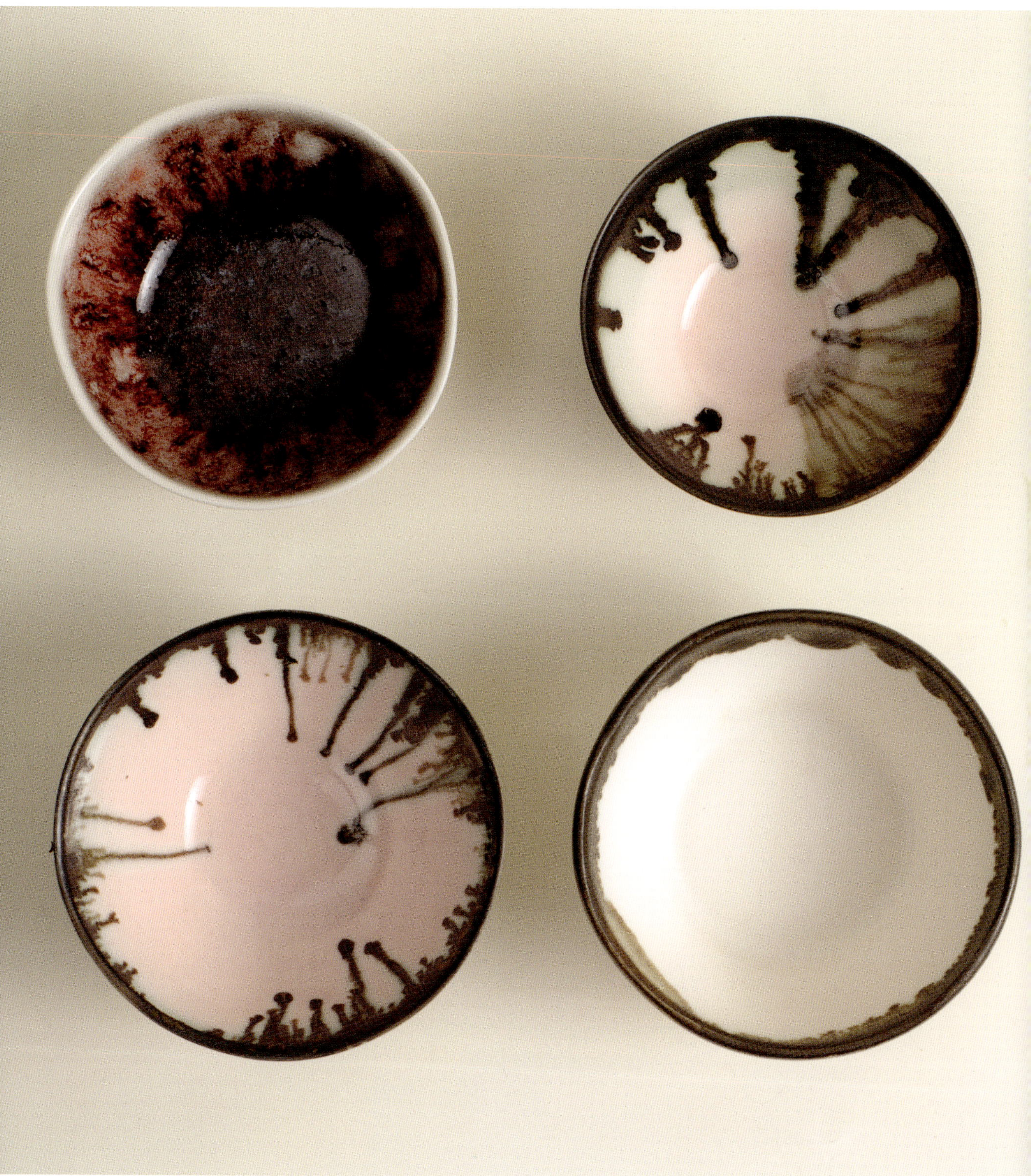

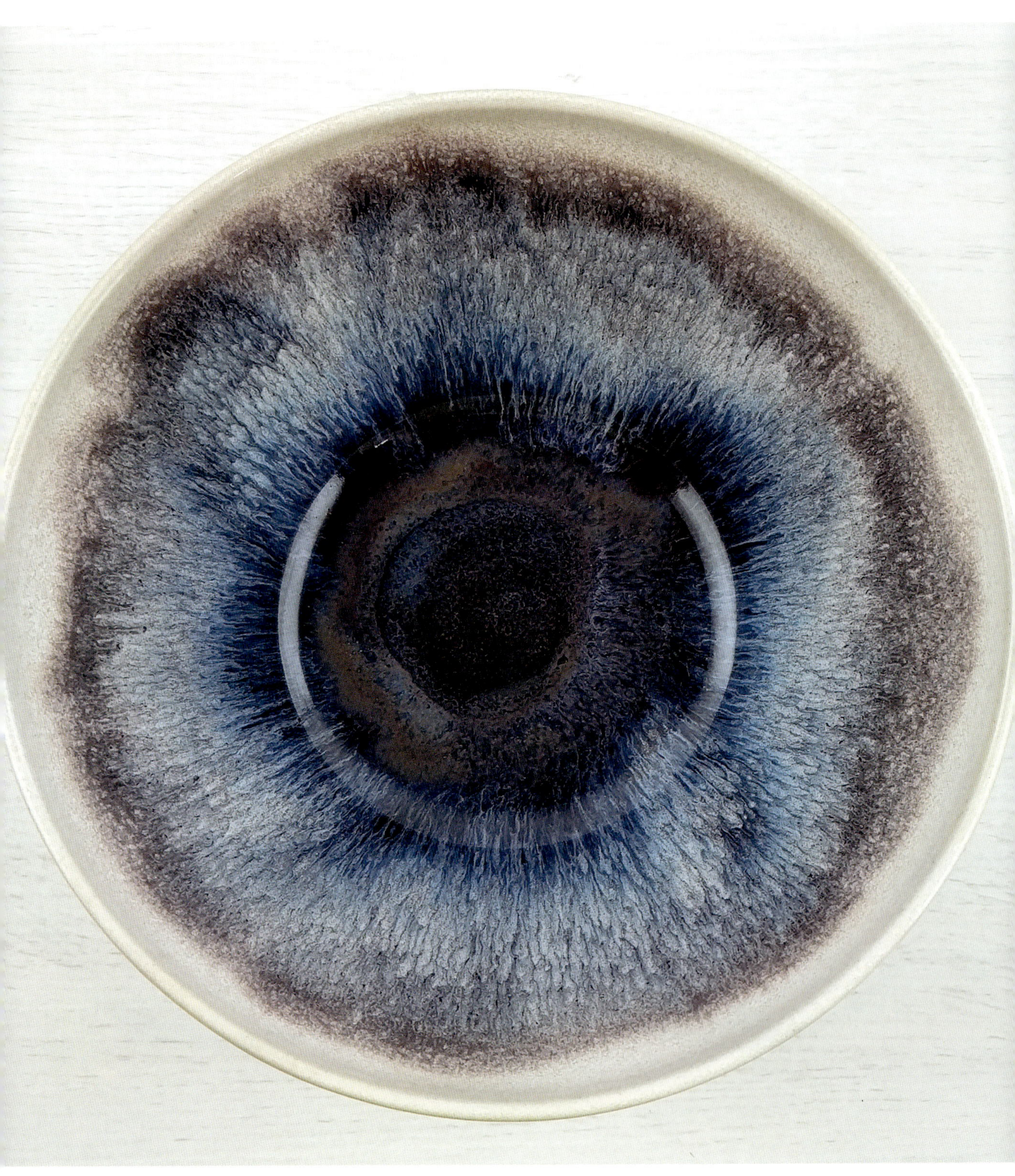

20 Layering glazes

Two different glazes can be applied in layers, which may react together. Alternatively, a reactive slip or engobe containing silicon carbide can be applied under a matt glaze. Some potters apply the slip to leatherhard or bone-dry ware, while others apply an engobe or biscuit slip at the biscuit stage. Colouring oxides such as iron oxide, manganese dioxide or granular ilmenite can be added to the clay body or slip and will cause pinholes, craters and dark spots to form in the glaze.

LEFT: Joe Thompson, Old Forge Creations, thrown stoneware bowl, June Perry Purple (chrome-tin) glaze over Storm Blue (cobalt rutile) glaze combination, fired to cone 5–6, dia: 20cm (8in.), height: 8cm (3⅛in.).

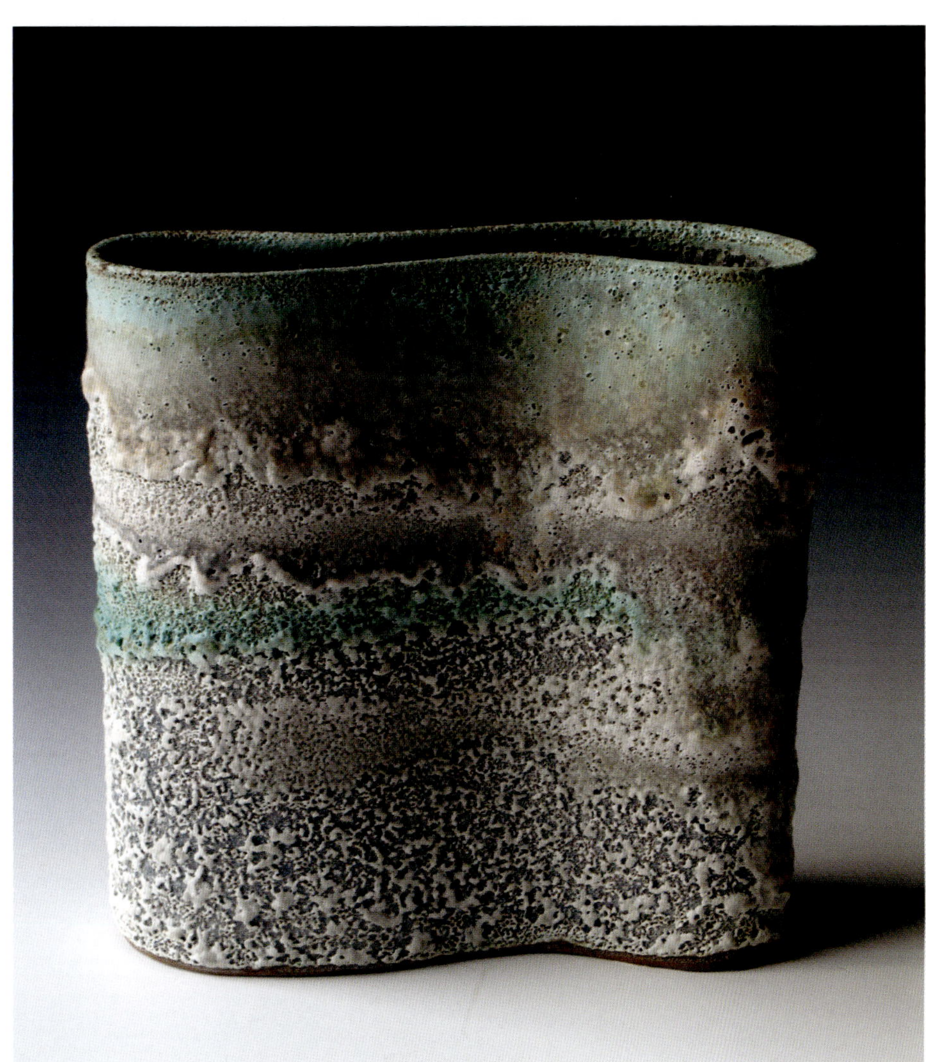

RIGHT: Katrina Pechal, stoneware vase with biscuit slip containing silicon carbide, barium glazes containing copper and vanadium, fired to cone 9.

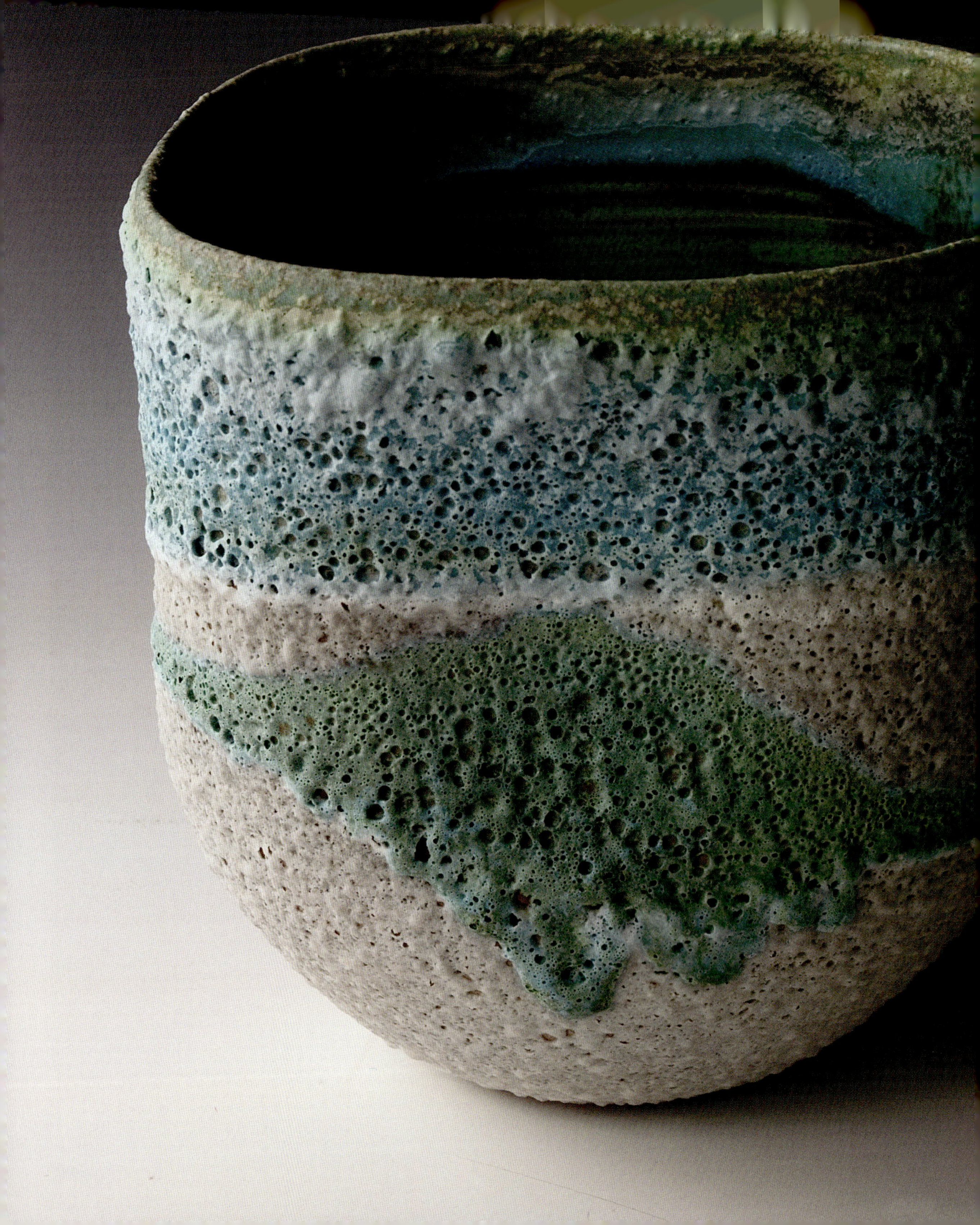

Conclusion

Special-effect glazes are often on the limits of what makes a stable, durable glaze. Crystalline glazes can be low in alumina, as can drippy glazes. Crackle glazes can be made using materials that have a high coefficient of expansion, such as nepheline syenite, while crawl glazes use materials that shrink on drying, such as light magnesium carbonate, clay or zinc oxide. Materials that release gases during firing can be used to generate a range of effects, from oil-spot to volcanic crater glazes. Silicon carbide has another use in its finely ground state, as a reducing agent, enabling potters to make celadon and copper red glazes in an electric kiln. Metallic effects can be made using an excess of colouring oxides such as manganese and copper oxides. These glazes are best used on decorative and sculptural pieces.

What I hope you can take from this book is how to manipulate 'normal' glazes to make various special effects. I have set out some of the ways you can push glazes outside normal limits to achieve interesting effects, and for each effect have given an insight into the mechanism of the chemistry and materials science that is behind it. I hope this has given you the confidence to try some of these special effects for yourself.

About the author

Linda Bloomfield is a scientist turned potter. More glaze recipes can be found in her books, *Advanced Pottery* (Robert Hale 2011), *Colour in Glazes* (A&C Black 2012, second edition Herbert Press 2019), *The Handbook of Glaze Recipes* (Bloomsbury 2014) and *Science for Potters* (The American Ceramic Society 2017).

Katrina Pechal, stoneware bowl with biscuit slip containing silicon carbide, barium glazes containing copper and vanadium, fired to cone 9.

References

Bloomfield, Linda, March 2014, 'Opalescent Chun-style glazes', *Ceramic Review*.

Bloomfield, Linda, January 2018, 'A material of many colours: rutile', *Ceramic Review*.

Bloomfield, Linda, 4, June 2018, 'Making glazes', *Clay Craft Magazine*.

Glazy.org, online glaze calculation by Derek Au.

Hansen, Tony, Ceramic materials online database, Digitalfire corporation, digitalfire.com.

Katz M., Gebhart T. and Carty W., 2003, 'The re-evaluation of the unity molecular formula limits for glazes', *Ceramic Engineering and Science Proceedings 24 (2)*, 13.

Katz, Matt, 2016, 'Glossed over: Durable Glazes', NCECA vol. 37.

Katz, Matt, Understanding Glazes lectures, CeramicMaterialsWorkshop.com.

Slade R. and Wood N., 2002, 'The production of classic Chinese glazes in oxidising kiln atmospheres using elemental silicon as a reducing agent'. Shanghai Institute of Ceramics, Chinese Academy of Sciences.

Stull, R.T., 1912, 'Influence of silica and alumina on porcelain glazes', *Transactions of the American Ceramic Society 14*, 62.

Bibliography

Bailey, Michael, *Glazes Cone 6, 1240°C*, A&C Black, 2001.

Bloomfield, Linda, *The Handbook of Glaze Recipes*, Bloomsbury, 2014.

Bloomfield, Linda, *Science for Potters*, 2017, American Ceramic Society.

Britt, John, *The Complete Guide to Mid-range Glazes: Glazing & Firing at Cones 4-7*, Lark Books, 2014.

Constant, Christine and Ogden, Steve, *The Potter's Palette*, Quarto, 1996.

Cooper, Emmanuel, *Cooper's Book of Glaze Recipes*, Batsford, 1987.

Cooper, Emmanuel, *The Complete Potter: Glazes*, Batsford, 1992.

Cooper, Emmanuel and Royle, Derek, *Glazes for the Studio Potter*, Batsford, 1984.

Currie, Ian, *Stoneware Glazes. A Systematic Approach*, Bootstrap Press, 1985.

Currie, Ian, *Revealing Glazes. Using the Grid Method*, Bootstrap Press, 2000.

Daly, Greg, *Developing Glazes*, Bloomsbury, 2013

De Montmollin, Daniel, 'The practice of stoneware glazes, minerals, rocks, ashes', *La Revue de La Céramique at Du Verre*, France, 2005.

Forrest, Miranda, *Natural Glazes: collecting and making*, A&C Black, 2013.

Fraser, Harry, *Glazes for the Craft Potter*, A&C Black, 1973.

Green, D., *Pottery Glazing Basics*, Coles Publishing Company, 1980.

Hamer, F. & J., *The Potter's Dictionary of Materials and Techniques*, sixth edition, Bloomsbury, 2015.

Hesselberth, John and Roy, Ron, *Mastering Cone 6 Glazes: Improving Durability, Fit and Aesthetics*, Glaze Master Press, 2002.

Hopper, Robin, *The Ceramic Spectrum: A simplified approach to glaze and colour development*, Krause publications, 1984.

Jernegan, Jeremy, *Dry Glazes*, A&C Black, 2009

Murfitt, Stephen, *The Glaze Book*, Thames and Hudson, 2002.

Parmelee, C.W. *Ceramic Glazes*, revised by C.G. Harman, third edition, Cahners publishing company, 1973.

Rhodes, Daniel, *Clay and Glazes for the Potter*, Krause publications, 1973.

Rogers, Phil, *Ash Glazes*, A&C Black, 1991.

Sanders, Herbert H., *Glazes for Special Effects*, Watson-Guptill, 1975.

Taylor, J.R. and Bull, A.C., *Ceramics Glaze Technology*, Pergamon Press, 1986.

Appendices

APPENDIX 1 UK:US materials substitutions

Some materials have no direct equivalent, but they can be substituted with a combination of several other materials.

UK	*USA*
Ball clay HVAR	Tennessee ball clay
Ball clay Hymod AT	Kentucky OM-4 ball clay
Ball clay Hyplas 71	Kentucky stone
Bentonite	Bentonite, Macaloid, Veegum
Borax frit	Ferro 3124, Pemco P-54, Gerstley borate
Calcium borate frit	Ferro 3134, Ferro 3195, Colemanite
China clay	Edgar plastic kaolin (EPK), Tile 6 kaolin, Georgia kaolin
Cornish Stone	Cornwall stone
Dispex	Darvan
Feldspar FFF	Custer feldspar plus Minspar or nepheline syenite
Feldspar potash	Custer feldspar, G-200, Mahavir feldspar
Feldspar soda	Kona F-4, NC 4, Minspar 200
Flint	Silica
Fremington red clay	Redart clay, Alberta slip
High-alkaline frit	Ferro 3110, Pemco P-25
Low-expansion frit	Ferro 3249
Quartz	Silica
Zirconium silicate	Zircopax, Superpax, Ultrox

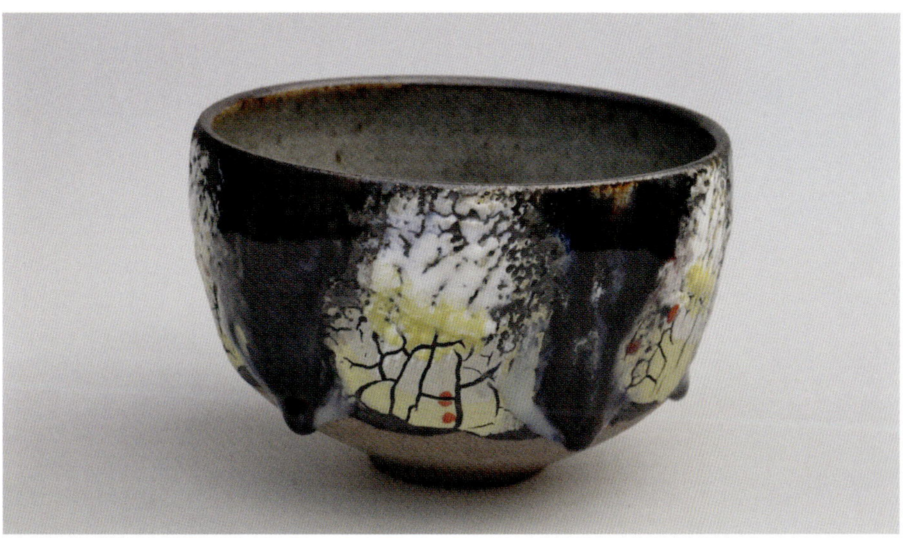

Mirka Golden-Hann, Earth Series stoneware bowl, vitreous crackle slips and glazes, reduction fired.

APPENDIX 2 Orton pyrometric cone temperatures

Pyrometric cones measure heatwork, so they depend on the heating rate. A slower temperature rise will cause the cone to bend at a lower temperature.

Cone no.	60°C/hour	108°F/hour	150°C/hour	270°F/hour
09	917	1683	928	1702
08	942	1728	954	1749
07	973	1783	984	1805
06	995	1823	985	1852
05	1030	1886	1046	1915
04	1060	1940	1070	1958
03	1086	1987	1101	2014
02	1101	2014	1120	2048
01	1117	2043	1137	2079
1	1136	2077	1154	2109
2	1142	2088	1162	2124
3	1152	2106	1168	2134
4	1160	2120	1181	2158
5	1184	2163	1205	2201
6	1220	2228	1241	2266
7	1237	2259	1255	2291
8	1247	2277	1269	2316
9	1257	2295	1278	2332
10	1282	2340	1303	2377
11	1293	2359	1312	2394
12	1304	2379	1324	2415
13	1321	2410	1346	2455
14	1388	2530	1366	2491

APPENDIX 3 Ceramic materials list

Ceramic materials, chemical formulae and molecular or equivalent* weights (containing one molecule).

Material	Formula	Molecular weight
Alumina	Al_2O_3	102
Alumina hydrate	$Al(OH)_3$	78*
Barium carbonate	$BaCO_3$	197.3
Bentonite	$Al_2O_3.4SiO_2.H_2O$	360.3
Bone ash (calcium phosphate)	$Ca_3(PO_4)_2$	103*
Borax	$Na_2O.B_2O_3.10\ H_2O$	381.4
Boric oxide	B_2O_3	69.6
Calcium borate	$Ca(BO2)_2$	125.7
Cerium oxide	CeO_2	172.1
China clay	$Al_2O_3.2SiO_2.2H_2O$	258.2
Colemanite	$2CaO.3B_2O_3.5H_2O$	206*
Cornwall stone	$K2O.Al_2O_3.8SiO_2$	676.8
Chromium oxide	Cr_2O_3	152
Cobalt oxide	CoO	74.9
Copper oxide	CuO	79.5
Cryolite	Na_3AlF_6	210
Dolomite	$CaCO_3.MgCO_3$	184.4
Erbium oxide	Er_2O_3	382.5
Feldspar soda (albite)	$Na_2O.Al_2O_3.6SiO_2$	524.4
Feldspar potash (orthoclase)	$K2O.Al_2O_3.6SiO_2$	556.4
Feldspar lime (anorthite)	$CaO.Al_2O_3.2SiO_2$	278.2
Fluorspar	CaF_2	78.1
Ilmenite	$FeO.TiO_2$	151.7
Iron (ferric) oxide (red)	Fe_2O_3	159.7
Iron (ferrous) oxide (black)	FeO	71.8
Kaolin	$Al_2O_3.2SiO_2.2H_2O$	258.2
Kyanite	$Al_2O_3.SiO_2$	162
Lead bisilicate	$PbO.2SiO_2$	343.4
Lead oxide (litharge)	PbO	223.2
Lead sulphide (galena)	PbS	239.2
Lead sesquisilicate	$2PbO.3SiO_2$	313.3*
Lepidolite	$LiFKF.Al_2O_3.3SiO_2$	366.3

Material	Formula	Molecular weight
Lithium carbonate	Li_2CO_3	73.9
Magnesia	MgO	40.3
Magnesium carbonate	Mg_2CO_3	84.3
Manganese dioxide	MnO_2	87
Mullite	$3Al_2O_3.2SiO_2$	426.1
Neodymium oxide	Nd_2O_3	336.5
Nepheline syenite	$K_2O.3Na_2O.4Al_2O_3.8SiO_2$	389.6*
Nickel oxide	NiO	74.7
Petalite	$Li_2O.Al_2O_3.8SiO_2$	612.5
Praseodymium oxide	Pr_2O_3	329.8
Quartz	SiO_2	60.1
Rutile	TiO_2	79.9
Silicon carbide	SiC	40.1
Silica	SiO_2	60.1
Spinel	$MgAl_2O_4$	142.3
Spodumene	$Li_2O.Al_2O_3.4SiO_2$	372.2
Strontium carbonate	$SrCO_3$	147.6
Talc (magnesium silicate)	$3MgO.4SiO_2.H_2O$	126.4*
Tin oxide	SnO_2	150.7
Titanium oxide (anatase)	TiO_2	79.9
Vanadium pentoxide	V_2O_5	181.9
Whiting (calcium carbonate)	$CaCO_3$	100.1
Wollastonite (calcium silicate)	$CaSiO_3$	116.2
Zinc oxide	ZnO	81.4
Zirconium silicate	$ZrSiO_4$	183.3

*Equivalent weights (containing one molecule) are sometimes given where the chemical formula contains a multiple number of molecules. For example, talc contains three molecules of magnesia, so the total molecular weight is divided by three to give the equivalent weight of talc containing one molecule of magnesia.

APPENDIX 4 Limits for stable glazes in the unity molecular formula

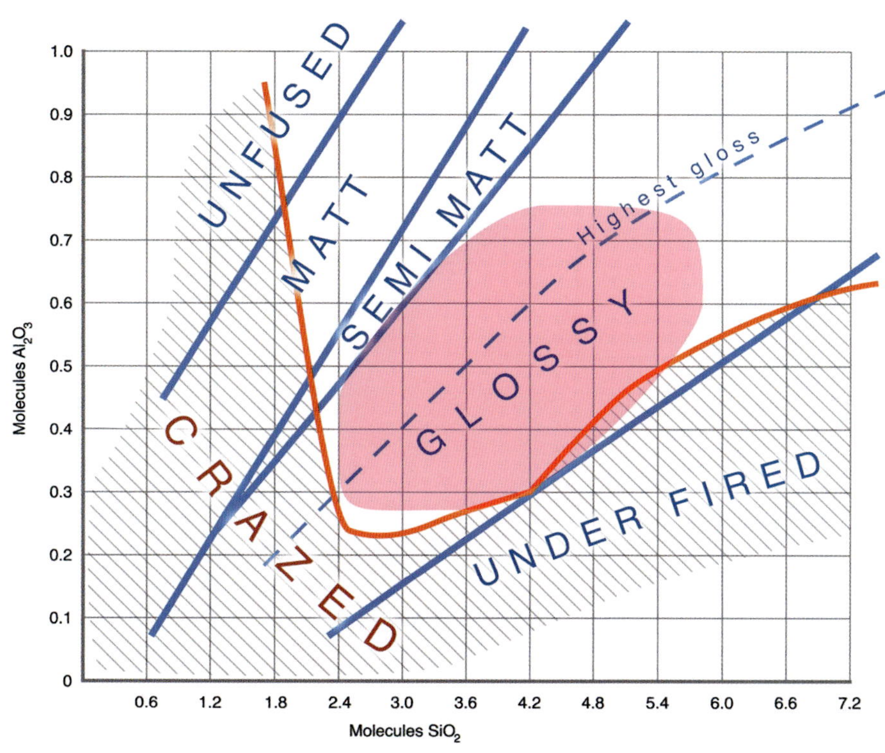

Stable glazes (those that are dishwasher-safe and food-safe) will generally have a ratio of alkali metal (potassium, sodium) to alkaline earth (calcium, magnesium) oxides of around 0.3:0.7 (±0.1). The amounts of silica and alumina will increase with firing temperature. For glazes fired below cone 9 (1280°C/2336°F), enough boron is needed to fully melt the glaze (0.1 B_2O_3 for cone 8 glazes to 0.5 B_2O_3 for cone 04 earthenware glazes).

Graph of alumina against silica in porcelain glazes fired to cone 11 with constant flux 0.3 K_2O and 0.7 CaO. The ratio of 1:5 alumina to silica gives a semi-matt glaze, while 1:8 gives a shiny glaze. The straight lines on the chart represent alumina:silica ratios of 1:4 (matt), 1:5 (semi-matt) and 1:12 (glossy, crazed glaze). The dashed line is 1:8 Al_2O_3:SiO_2 (highest gloss glaze). The hatched area shows crazed glazes on porcelain. The pink area shows glaze limits for cones 5–8 (Cooper and Royle 1984). Data from R.T. Stull 1912.

Alumina and silica limits (Cooper and Royle, 1984).

Cone number and temperature.		Number of molecules in unity formula.			
Cone 04	1060°C (1940°F)	Al_2O_3	0.1–0.45	SiO_2	1.375–3.15
Cone 5	1200°C (2192°F)	Al_2O_3	0.275–0.65	SiO_2	2.4–4.7
Cone 6	1225°C (2237°F)	Al_2O_3	0.325–0.70	SiO_2	2.6–5.15
Cone 8	1250°C (2282°F)	Al_2O_3	0.375–0.75	SiO_2	3.0–5.75
Cone 9	1275°C (2327°F)	Al_2O_3	0.45–0.825	SiO_2	3.5–6.4
Cone 10	1300 °C (2372°F)	Al_2O_3	0.50–0.90	SiO_2	4.0–7.2

Recommended maximum flux in molecular unity formula (Cooper and Royle, 1984).

Cone	Temp °C (°F)	MgO	BaO	ZnO	CaO	B2O3	K+Na
5	1200°C (2192°F)	0.325	0.40	0.30	0.55	0.35	0.375
6	1225°C (2237°F)	0.330	0.425	0.32	0.60	0.30	0.35
8	1250°C (2282°F)	0.335	0.45	0.34	0.65	0.25	0.325
9	1275°C (2327°F)	0.340	0.475	0.36	0.70	0.225	0.30
10	1300 °C (2372°F)	0.345	0.50	0.38	0.75	0.21	0.275

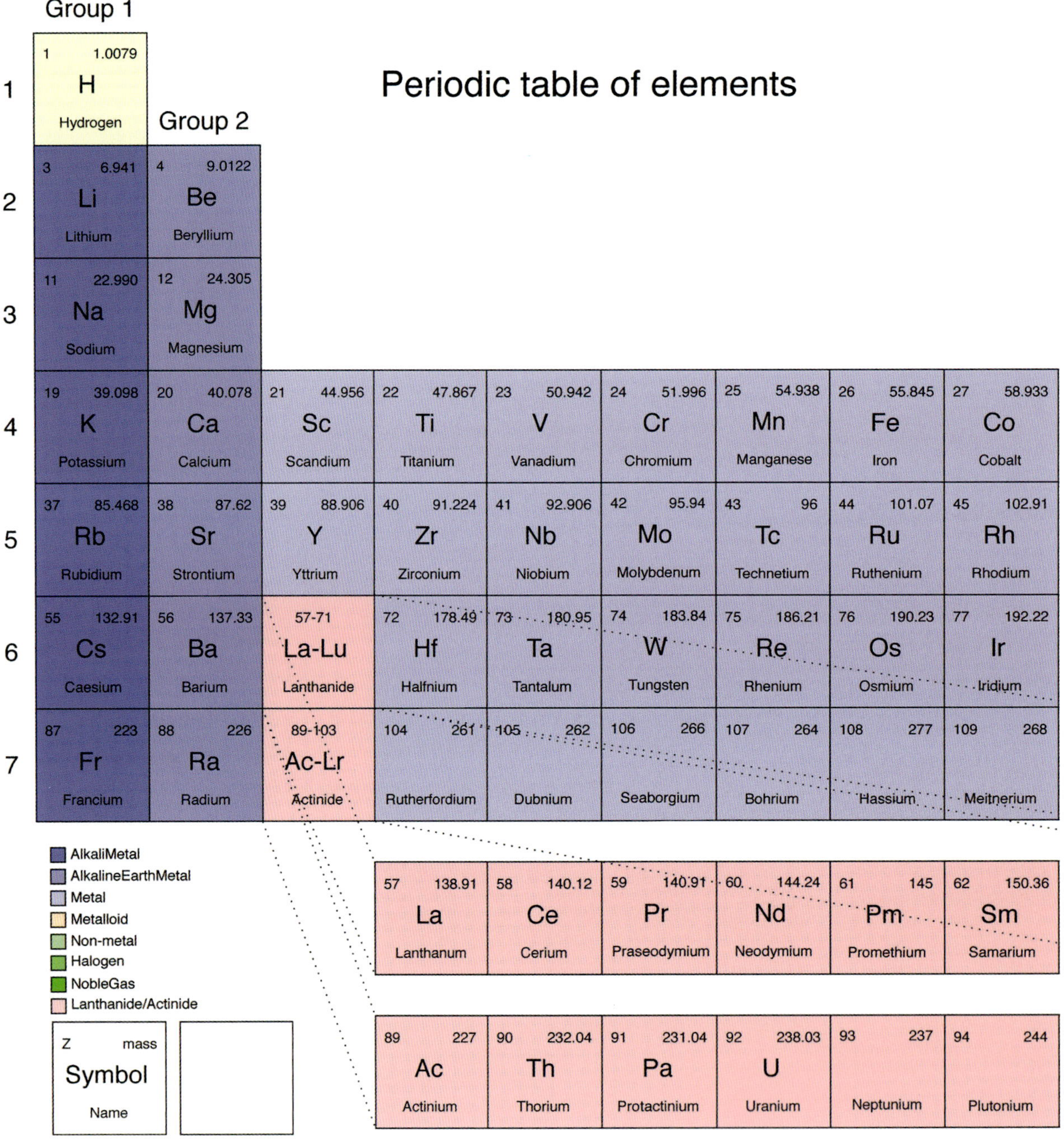
Periodic table of elements
Group 1
Group 2
1
2
3
4
5
6
7
1 1.0079 H Hydrogen
3 6.941 Li Lithium
4 9.0122 Be Beryllium
11 22.990 Na Sodium
12 24.305 Mg Magnesium
19 39.098 K Potassium
20 40.078 Ca Calcium
21 44.956 Sc Scandium
22 47.867 Ti Titanium
23 50.942 V Vanadium
24 51.996 Cr Chromium
25 54.938 Mn Manganese
26 55.845 Fe Iron
27 58.933 Co Cobalt
37 85.468 Rb Rubidium
38 87.62 Sr Strontium
39 88.906 Y Yttrium
40 91.224 Zr Zirconium
41 92.906 Nb Niobium
42 95.94 Mo Molybdenum
43 96 Tc Technetium
44 101.07 Ru Ruthenium
45 102.91 Rh Rhodium
55 132.91 Cs Caesium
56 137.33 Ba Barium
57-71 La-Lu Lanthanide
72 178.49 Hf Halfnium
73 180.95 Ta Tantalum
74 183.84 W Tungsten
75 186.21 Re Rhenium
76 190.23 Os Osmium
77 192.22 Ir Iridium
87 223 Fr Francium
88 226 Ra Radium
89-103 Ac-Lr Actinide
104 261 Rutherfordium
105 262 Dubnium
106 266 Seaborgium
107 264 Bohrium
108 277 Hassium
109 268 Meitnerium
AlkaliMetal
AlkalineEarthMetal
Metal
Metalloid
Non-metal
Halogen
NobleGas
Lanthanide/Actinide
Z mass Symbol Name
57 138.91 La Lanthanum
58 140.12 Ce Cerium
59 140.91 Pr Praseodymium
60 144.24 Nd Neodymium
61 145 Pm Promethium
62 150.36 Sm Samarium
89 227 Ac Actinium
90 232.04 Th Thorium
91 231.04 Pa Protactinium
92 238.03 U Uranium
93 237 Neptunium
94 244 Plutonium

			Group 3	Group 4	Group 5	Group 6	Group 7	Group 8
								2 4.0025 He Helium
			5 10.811 B Boron	6 12.011 C Carbon	7 14.007 N Nitrogen	8 15.999 O Oxygen	9 18.998 F Flourine	10 20.180 Ne Neon
			13 26.982 Al Aluminium	14 28.086 Si Silicon	15 30.974 P Phosphorus	16 32.065 S Sulphur	17 35.453 Cl Chlorine	18 39.948 Ar Argon
28 58.693 Ni Nickel	29 63.546 Cu Copper	30 65.39 Zn Zinc	31 69.723 Ga Gallium	32 72.64 Ge Germanium	33 74.922 As Arsenic	34 78.96 Se Selenium	35 79.904 Br Bromine	36 83.8 Kr Krypton
46 106.42 Pd Palladium	47 107.87 Ag Silver	48 112.41 Cd Cadmium	49 114.82 In Indium	50 118.71 Sn Tin	51 121.76 Sb Antimony	52 127.6 Te Tellurium	53 126.9 I Iodine	54 131.29 Xe Xenon
78 195.08 Pt Platinum	79 196.97 Au Gold	80 200.59 Hg Mercury	81 204.38 Tl Thallium	82 207.2 Pb Lead	83 208.98 Bi Bismuth	84 209 Po Polonium	85 210 At Astatine	86 222 Rn Radon
110 281 Darmstadtium	111 280 Roentgenium	112 285 Copernicum	113 284 Nihonium	114 289 Flerovium	115 288 Moscovium	116 293 Livermorium	117 292 Tennessine	118 294 Oganesson

63 151.96 Eu Europium	64 157.25 Gd Gadolinium	65 158.93 Tb Terbium	66 162.50 Dy Dysprosium	67 164.93 Ho Holmium	68 167.26 Er Erbium	69 168.93 Tm Thulium	70 173.04 Yb Ytterbium	71 174.97 Lu Lutetium

95 243 Americium	96 247 Curium	97 247 Berkelium	98 251 Californium	99 252 Einsteinium	100 257 Fermium	101 258 Mendelevium	102 259 Nobelium	103 262 Lawrencium

APPENDIX 5 Materials analysis for UK frits, clays and feldspars

Frits, unity formula analysis (Bath Potters Supplies, Michael Bailey)

	K_2O	Na_2O	Li_2O	BaO	CaO	MgO	ZnO	Al_2O_3	B_2O_3	SiO_2	Mol. wt.
Calcium borate frit	0.01				0.99	0.01		0.1	1.5	0.62	209
Standard borax frit	0.04	0.35			0.61	0.01		0.18	0.62	1.98	240
High-alkaline frit	0.21	0.59	0.01	0.09	0.1			0.1	0.1	1.71	196
Low-expansion frit	0.03	0.2			0.76	0.01		0.55	1.02	3.39	390

Clays and feldspars, percentage composition (LOI = loss on ignition)

	SiO_2	TiO_2	Al_2O_3	Fe_2O_3	P_2O_5	CaO	MgO	K_2O	Na_2O	LOI
China clay	48.8	0.1	35.4	0.8				1.6	1.5	11.8
AT Ball clay	54	1.1	29	2.4		0.3	0.4	3	0.5	9.3
HP71 Ball clay	70	1.6	19	0.8		0.2	0.4	2	0.5	5.5
HVAR Ball clay	60.3	1.5	26.7	0.9		0.2	0.3	2.6	0.4	7.1
Cornish stone	73.2	0.06	15.3	0.13	0.47	1.47	0.13	4.45	3.44	1.35
Nepheline syenite	60.5		23	0.1		1		5	10.2	0.2
Potash feldspar	65.8		18.5	0.1		0.38		12	2.89	0.33
Soda feldspar	67.9		19	0.11		1.88		2.8	7.5	0.81
FFF feldspar	67.7		18.9	0.16		0.72		7.62	4.85	0.05

APPENDIX 6 Materials analysis for US frits, clays and feldspars

Frits, unity formula analysis (Michael Bailey)

	K_2O	Na_2O	Li_2O	BaO	CaO	MgO	ZnO	Al_2O_3	B_2O_3	SiO_2	Mol. wt.
Ferro 3110	0.060	0.650			0.290			0.095	0.097	3.029	260
Ferro 3124	0.021	0.282			0.698			0.269	0.519	2.554	275
Ferro 3134		0.317			0.683				0.633	1.890	191
Ferro 3195		0.336			0.654	0.01		0.392	1.136	2.656	337
Ferro 3249					0.171	0.829		0.357	1.137	1.919	274

Clays and feldspars, percentage composition (LOI = loss on ignition)

	SiO_2	TiO_2	Al_2O_3	Fe_2O_3	P_2O_5	CaO	MgO	K_2O	Na_2O	LOI
EPK	45.91	0.34	38.71	0.42		0.09	0.12	0.22	0.04	14.15
Georgia kaolin (Tile #6)	45.20	1.95	38.02	0.49		0.26	0.30	0.04	0.02	13.72
OM-4 Ball clay	55.20	1.20	29.70	1.10		0.30	0.40	1.00	0.30	12.60
Tennessee no.5 Ball clay	53.30	1.40	31.10	1.00		0.30	0.20	1.50	0.80	10.40
Custer feldspar	68.50		17.50	0.08		0.30	0.01	10.40	3.00	0.21
G-200 feldspar	65.76		19.28	0.06		0.98	0.01	10.36	3.20	0.35
Kona F-4 feldspar	66.77		19.59	0.04		1.70	0.01	4.50	7.00	0.39
NC-4 feldspar	68.81		18.74	0.07		1.60	0.01	3.76	6.89	0.12
Minspar 200	68.80		18.20	0.07		1.50		4.10	6.50	0.30
Mahavir feldspar	67.00		17.50	0.08		0.15	0.15	11.5	3.00	0.50

Suppliers

UK

Bath Potters' Supplies
Unit 18, Fourth Avenue,
Westfield Trading Estate,
Radstock, Nr Bath,
Somerset,
BA3 4XE
Tel: +44 (0)1761 411077
www.bathpotters.co.uk

Ceramatech
16–17 Frontier Works,
33 Queen Street,
London N17 8JA
Tel: +44 (0) 208 885 4492
www.ceramatech.co.uk

Clayman
Morrells Barn,
Lower Bognor Road,
Lagness, Chichester, West Sussex,
PO20 1LR
Tel: 01243 265845
www.claymansupplies.co.uk

CTM Potters Supplies
Unit 10A, Mill Park Industrial Estate,
White Cross Road,
Woodbury Salterton,
Exeter,
Devon,
EX5 1EL
Tel: +44 (0)1395 233077
www.ctmpotterssupplies.co.uk

Potclays
Brickkiln Lane, Etruria,
Stoke-on-Trent,
Staffordshire,
ST4 7BP
Tel: +44 (0)1782 219816
www.potclays.co.uk

Potterycrafts
Campbell Road,
Stoke-on-Trent,
Staffordshire,
ST4 4ET
Tel: +44 (0)1782 745000
www.potterycrafts.co.uk

Scarva Pottery Supplies
Unit 20,
Scarva Road Industrial Estate,
Banbridge, County Down,
Northern Ireland,
BT32 3QD
Tel: +44 (0)28 406 69699
www.scarvapottery.co.uk

Spencroft Ceramics
Holditch Industrial Estate,
Spencroft Road,
Stoke-on-Trent,
Staffordshire,
ST5 9JB
Tel: +44 (0)1782 717305
www.spencroftceramics.co.uk

Valentine Clays
The Sliphouse,
18–20 Chell Street,
Hanley,
Stoke-on-Trent,
Staffordshire,
ST1 6BA
Tel: +44 (0)1782 271200
www.valentineclays.co.uk

USA

Amaco Brent
6060 Guion Road,
Indianapolis,
IN 46254-1222
Tel: +1 (317) 244 6871
www.amaco.com

Bailey Ceramic Supplies
62–68 Tenbroeck Avenue,
Kingston,
New York 12401
Tel: +1 (845) 339 3721
www.baileypottery.com

Bracker's Good Earth Clays,
1831 E. 1450 Road
Lawrence, KS 66044
Tel: +1 (785) 841 4750
www.brackers.com

Clay Planet
1775 Russell Avenue,
Santa Clara, CA 95054
Tel: +1 (800) 443 2529
www.clay-planet.com

Columbus Clay Co.
1080 Chambers Road,
Columbus, OH 43212
Tel: +1 (866) 410 2529
www.columbusclay.com

Georgie's Ceramic & Clay Co
756 NE Lombard
Portland, OR 97211
Tel 503 283 1353
www.georgies.com

Hamill and Gillespie
420 Clermont Terrace
Unit D, Union,
NJ 07083
Tel: 973 822 8000
www.hamgil.com

Laguna Clay Co.
14400 Lomitas Avenue,
City of Industry, CA 91746
Tel: +1 (800) 452 4862
www.lagunaclay.com

Minnesota Clay Co
2960 Niagara Lane
Plymouth,
MN 55447
Tel: 763 432 0875
www.mnclay.com

US pigments
815 Schneider Drive
South Elgin,
IL 60177
Tel: 630 893 9217
uspigment.com

Canada

Plainsman Clays
702 Wood Street SE,
Medicine Hat,
Alberta,
Canada,
T1A 1E9
Tel: 403-527-8535
plainsmanclays.com

Australia

Pottery Supplies Brisbane
51 Castlemaine Street,
Milton ,
Queensland 4064
Tel:07 3368 2877
www.potterysupplies.com.au

Walker Ceramics
5 McLellan Street,
Bayswater,
Victoria 3153
Tel: 03 8761 6322
www.walkerceramics.com.au

Laboratories for leach-testing of glazes

Lucideon
Queens Road,
Penkhull,
Stoke-on-Trent,
Staffordshire,
ST4 7LQ,
UK
01782 764428
lucideon.com

Northern Testhouse,
Scraptoft Business Centre
Main Street,
Scraptoft,
Leicester LE7 9TD
UK
01162 418811
nthleicester.co.uk

Brandywine Science Center
204 Line Road,
Kennet Square,
PA 19348,
USA
www.bsclab.com

Health and safety

- It is important to reduce dust in the studio by cleaning up often and avoiding build-up of dried clay or glaze on any surface: work bench, clothes or floor.
- Use a respirator face mask when weighing dry powdered materials, particularly those containing silica, including quartz, flint, talc, wollastonite, feldspar, clay and frits.
- Clean up spills straight away and wash surfaces and floors using a wet sponge or mop.
- Wash towels, aprons and overalls regularly.
- Never drink or eat in the studio.
- Dry out and dispose of waste glazes in landfill, do not wash down the drain. Better still, mix all the waste glazes together in one bucket and use as a new glaze.
- Make sure the kiln is well ventilated and do not breathe in the fumes.

Index